SPARSE MODELING

Theory, Algorithms, and Applications

Chapman & Hall/CRC
Machine Learning & Pattern Recognition Series

SERIES EDITORS

Ralf Herbrich
Amazon Development Center
Berlin, Germany

Thore Graepel
Microsoft Research Ltd.
Cambridge, UK

AIMS AND SCOPE

This series reflects the latest advances and applications in machine learning and pattern recognition through the publication of a broad range of reference works, textbooks, and handbooks. The inclusion of concrete examples, applications, and methods is highly encouraged. The scope of the series includes, but is not limited to, titles in the areas of machine learning, pattern recognition, computational intelligence, robotics, computational/statistical learning theory, natural language processing, computer vision, game AI, game theory, neural networks, computational neuroscience, and other relevant topics, such as machine learning applied to bioinformatics or cognitive science, which might be proposed by potential contributors.

PUBLISHED TITLES

BAYESIAN PROGRAMMING
Pierre Bessière, Emmanuel Mazer, Juan-Manuel Ahuactzin, and Kamel Mekhnacha

UTILITY-BASED LEARNING FROM DATA
Craig Friedman and Sven Sandow

HANDBOOK OF NATURAL LANGUAGE PROCESSING, SECOND EDITION
Nitin Indurkhya and Fred J. Damerau

COST-SENSITIVE MACHINE LEARNING
Balaji Krishnapuram, Shipeng Yu, and Bharat Rao

COMPUTATIONAL TRUST MODELS AND MACHINE LEARNING
Xin Liu, Anwitaman Datta, and Ee-Peng Lim

MULTILINEAR SUBSPACE LEARNING: DIMENSIONALITY REDUCTION OF MULTIDIMENSIONAL DATA
Haiping Lu, Konstantinos N. Plataniotis, and Anastasios N. Venetsanopoulos

MACHINE LEARNING: An Algorithmic Perspective, Second Edition
Stephen Marsland

SPARSE MODELING: THEORY, ALGORITHMS, AND APPLICATIONS
Irina Rish and Genady Ya. Grabarnik

A FIRST COURSE IN MACHINE LEARNING
Simon Rogers and Mark Girolami

MULTI-LABEL DIMENSIONALITY REDUCTION
Liang Sun, Shuiwang Ji, and Jieping Ye

REGULARIZATION, OPTIMIZATION, KERNELS, AND SUPPORT VECTOR MACHINES
Johan A. K. Suykens, Marco Signoretto, and Andreas Argyriou

ENSEMBLE METHODS: FOUNDATIONS AND ALGORITHMS
Zhi-Hua Zhou

Chapman & Hall/CRC
Machine Learning & Pattern Recognition Series

SPARSE MODELING

Theory, Algorithms, and Applications

Irina Rish

IBM
Yorktown Heights, New York, USA

Genady Ya. Grabarnik

St. John's University
Queens, New York, USA

CRC Press
Taylor & Francis Group
Boca Raton   London   New York

CRC Press is an imprint of the
Taylor & Francis Group, an **informa** business

A CHAPMAN & HALL BOOK

CRC Press
Taylor & Francis Group
6000 Broken Sound Parkway NW, Suite 300
Boca Raton, FL 33487-2742

© 2015 by Taylor & Francis Group, LLC
CRC Press is an imprint of Taylor & Francis Group, an Informa business

No claim to original U.S. Government works

Printed on acid-free paper
Version Date: 20141017

International Standard Book Number-13: 978-1-4398-2869-4 (Hardback)

Visit the Taylor & Francis Web site at
http://www.taylorandfrancis.com

and the CRC Press Web site at
http://www.crcpress.com

To Mom, my brother Ilya, and my family – Natalie, Alexander, and Sergey. And in loving memory of my dad and my brother Dima.

To Fany, Yaacob, Laura, and Golda.

Contents

List of Figures

Preface

If Ptolemy, Agatha Christie, and William of Ockham had a chance to meet, they would probably agree on one common idea. "We consider it a good principle to explain the phenomena by the simplest hypothesis possible," Ptolemy would say. "The simplest explanation is always the most likely," Agatha would add. And William of Ockham would probably nod in agreement: "Pluralitas non est ponenda sine necesitate," i.e., "Entities should not be multiplied unnecessarily." This principle of parsimony, known today as Ockam's (or Occam's) razor, is arguable one of the most fundamental ideas that pervade philosophy, art and science from ancient to modern times. "Simplicity is the ultimate sophistication" (Leonardo da Vinci). "Make everything as simple as possible, but not simpler" (Albert Einstein). Endless quotes in favor of simplicity from many great minds in the history of humankind could easily fill out dozens of pages. But we would rather keep this preface short (and simple).

The topic of this book – *sparse modeling* – is a particular manifestation of the parsimony principle in the context of modern statistics, machine learning and signal processing. A fundamental problem in those fields is an accurate recovery of an unobserved high-dimensional signal from a relatively small number of measurements, due to measurement costs or other limitations. Image reconstruction, learning model parameters from data, diagnosing system failures or human diseases are just a few examples where this challenging inverse problem arises. In general, high-dimensional, small-sample inference is both underdetermined and computationally intractable, unless the problem has some specific structure, such as, for example, *sparsity*.

Indeed, quite frequently, the ground-truth solution can be well-approximated by a sparse vector, where only a few variables are truly important, while the remaining ones are zero or nearly-zero; in other words, a small number of most-relevant variables (causes, predictors, etc.) can be often sufficient for explaining a phenomenon of interest. More generally, even if the original problem specification does not yield a sparse solution, one can typically find a mapping to a new a coordinate system, or *dictionary*, which allows for such sparse representation. Thus, sparse structure appears to be an inherent property of many natural signals – and without such structure, understanding the world and adapting to it would be considerably more challenging.

In this book, we tried to provide a brief introduction to sparse modeling, including application examples, problem formulations that yield sparse solutions, algorithms for finding such solutions, as well as some recent theoretical results on sparse recovery. The material of this book is based on our tutorial presented several years ago at the ICML-2010 (International Conference on Machine Learning), as well as on a graduate-level course that we taught at the Columbia University in the spring semester of 2011.

We start chapter 1 with motivating examples and a high-level survey of key recent developments in sparse modeling. In chapter 2, we formulate optimization problems that involve commonly used sparsity-enforcing tools such as l_0- and l_1-norm constraints. Essential theoretical results are presented in chapters 3 and 4, while chapter 5 discusses several well-known algorithms for finding sparse solutions. Then, in chapters 6 and 7, we discuss a variety of sparse recovery problems that extend the basic formulation towards more sophisticated forms of structured sparsity and towards different loss functions, respectively. Chapter 8 discusses a particular class of sparse graphical models such as sparse Gaussian Markov Random Fields, a popular and fast-developing subarea of sparse modeling. Finally, chapter 9 is devoted to dictionary learning and sparse matrix factorizations.

Note that our book is by no means a complete survey of all recent sparsity-related developments; in fact, no single book can fully capture this incredibly fast-growing field. However, we hope that our book can serve as an introduction to the exciting new field of sparse modeling, and motivate the reader to continue learning about it beyond the scope of this book.

Finally, we would like to thank many people who contributed to this book in various ways. Irina would like to thank her colleagues at the IBM Watson Research Center – Chid Apte, Guillermo Cecchi, James Kozloski, Laxmi Parida, Charles Peck, Ravi Rao, Jeremy Rice, and Ajay Royyuru – for their encouragement and support during all these years, as well as many other collaborators and friends whose ideas helped to shape this book, including Narges Bani Asadi, Alina Beygelzimer, Melissa Carroll, Gaurav Chandalia, Jean Honorio, Natalia Odintsova, Dimitris Samaras, Katya Scheinberg and Ben Taskar. Ben passed away last year, but he will continue to live in our memories and in his brilliant work.

The authors are grateful to Dmitry Malioutov, Aurelie Lozano, and Francisco Pereira for reading the manuscript and providing many valuable comments that helped to improve this book. Special thanks to Randi Cohen, our editor, for keeping us motivated and waiting patiently for this book to be completed. Last, but not least, we would like to thank our families for their love, support and patience, and for being our limitless source of inspiration. We have to admit that it took us a bit longer than previously anticipated to finish this book (only a few more years); as a result, Irina (gladly) lost a bet to her daughter Natalie about who will first publish a book.

Chapter 1

Introduction

A common question arising in a wide variety of practical applications is how to infer an unobserved high-dimensional "state of the world" from a limited number of observations. Examples include finding a subset of genes responsible for a disease, localizing brain areas associated with a mental state, diagnosing performance bottlenecks in a large-scale distributed computer system, reconstructing high-quality images from a compressed set of measurements, and, more generally, decoding any kind of signal from its noisy encoding, or estimating model parameters in a high-dimensional but small-sample statistical setting.

The underlying inference problem is illustrated in Figure 1.1, where $\mathbf{x} = (x_1, ..., x_n)$ and $\mathbf{y} = (y_1, ..., y_m)$ represent an n-dimensional unobserved state of the world and its m observations, respectively. The output vector of observations, $\mathbf{y}$, can be viewed as a noisy function (encoding) of the input vector $\mathbf{x}$. A commonly used inference (decoding) approach is to find $\mathbf{x}$ that minimizes some *loss function* $L(\mathbf{x}; \mathbf{y})$, given the observed $\mathbf{y}$. For example, a popular probabilistic *maximum likelihood* approach aims at finding a parameter vector $\mathbf{x}$ that maximizes the likelihood $P(\mathbf{y}|\mathbf{x})$ of the observations, i.e., minimizes the negative log-likelihood loss.

However, in many real-life problems, the number of unobserved variables greatly exceeds the number of measurements, since the latter may be expensive and also limited by the problem-specific constraints. For example, in computer network diagnosis, gene network analysis, and neuroimaging applications the total number of unknowns, such as states of network elements, genes, or brain voxels, can be on the order of thousands, or even hundreds of thousands, while the number of observations, or samples, is typically on the order of hundreds. Therefore, the above maximum-likelihood formulation becomes underdetermined, and additional *regularization* constraints, reflecting specific domain properties or assumptions, must be introduced in order to restrict the space of possible solutions. From a Bayesian probabilistic perspective, regularization can be viewed as imposing a *prior* $P(\mathbf{x})$ on the unknown

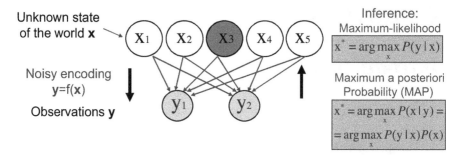

FIGURE 1.1: Is it possible to recover an unobserved high-dimensional signal **x** from a low-dimensional, noisy observation **y**? Surprisingly, the answer is positive, provided that **x** has some specific structure, such as (sufficient) sparsity, and the mapping $\mathbf{y} = f(\mathbf{x})$ preserves enough information in order to reconstruct **x**.

parameters **x**, and maximizing the posterior probability $P(\mathbf{x}|\mathbf{y}) = P(\mathbf{y}|\mathbf{x})P(\mathbf{x})/P(\mathbf{y})$, as we discuss in the next chapter.

Perhaps one of the simplest and most popular assumptions made about the problem's structure is the solution *sparsity*. In other words, it is assumed that only a relatively small subset of variables is truly important in a specific context: e.g., usually only a small number of simultaneous faults occurs in a system; a small number of nonzero Fourier coefficients is sufficient for an accurate representation of various signal types; often, a small number of predictive variables (e.g., genes) is most relevant to the response variable (a disease, or a trait), and is sufficient for learning an accurate predictive model. In all these examples, the solution we seek can be viewed as a sparse high-dimensional vector with only a few nonzero coordinates. This assumption aligns with a philosophical principle of parsimony, commonly referred to as *Occam's razor*, or *Ockham's razor*, and attributed to William of Ockham, a famous medieval philosopher, though it can be traced back perhaps even further, to Aristotle and Ptolemy. Post-Ockham formulations of the principle of parsimony include, among many others, the famous one by Isaac Newton: "We are to admit no more causes of natural things than such as are both true and sufficient to explain their appearances".

Statistical models that incorporate the parsimony assumption will be referred to as *sparse models*. These models are particularly useful in scientific applications, such as biomarker discovery in genetic or neuroimaging data, where the interpretability of a predictive model, e.g., identification of the most-relevant predictors, is essential. Another important area that can benefit from sparsity is signal processing, where the goal is to minimize signal acquisition costs while achieving high reconstruction accuracy; as we discuss later, exploiting sparsity can dramatically improve cost-efficiency of signal processing.

From a historical perspective, sparse signal recovery problem formulations can be traced back to 1943, or possibly even earlier, when the *combinatorial group testing* problem was first introduced in (Dorfman, 1943). The original motivation behind this problem was to design an efficient testing scheme using blood samples obtained from a large population (e.g., on the order of 100,000 people) in order to identify a relatively small number of infected people (e.g., on the order of 10). While testing each individual was considered prohibitively costly, one could combine the blood samples from groups of people; testing such combined samples would reveal if at least one person in the group had a disease. Following the inference scheme in Figure 1.1, one can represent the health state of the i-th person as a Boolean variable $\mathbf{x}_i$, where $x_i = 0$ if the person is healthy, and $x_i = 1$ otherwise; the test result, or measurement, y_j for a group of people G_j is the *logical-OR* function over the variables in the group, i.e. $y_j = 0$ if and only if all $x_i = 0, i \in G_j$, and 1 otherwise. Given an upper bound on the number of sick individuals in the population, i.e. the bound on sparsity of $\mathbf{x}$, the objective of group testing is to identify all sick individuals (i.e., nonzero x_i), while minimizing the number of tests.

Similar problem formulations arise in many other diagnostic applications, for example, in computer network fault diagnosis, where the network nodes, such as routers or links, can be either functional or faulty, and where the group tests correspond to end-to-end transactions, called network probes, that go through particular subsets of elements as determined by a routing table (Rish et al., 2005). (In the next section, we consider the network diagnosis problem in more detail, focusing, however, on its continuous rather than Boolean version, where the "hard faults" will be relaxed into performance bottlenecks, or time delays.) In general, group testing has a long history of successful applications to various practical problems, including DNA library screening, multiple access control protocols, and data streams, just to name a few. For more details on group testing, see the classical monograph by (Du and Hwang, 2000) and references therein, as well as various recent publications, such as, for example, (Gilbert and Strauss, 2007; Atia and Saligrama, 2012; Gilbert et al., 2012).

During the past several decades, half a century since the emergence of the combinatorial group testing field, sparse signal recovery is experiencing a new wave of intense interest, now with the primary focus on continuous signals and observations, and with particular ways of enforcing sparsity, such as using l_1-norm regularization. For example, in 1986, (Santosa and Symes, 1986) proposed an l_1-norm based optimization approach for the linear inversion (deconvolution) of band-limited reflection seismograms. In 1992, (Rudin et al., 1992) proposed *total variation* regularizer, which is closely related to l_1-norm, for noise removal in image processing. In 1996, the seminal paper by (Tibshirani, 1996) on LASSO, or the l_1-norm regularized linear regression, appeared in statistical literature, and initiated today's mainstream application of sparse regression to a wide range of practical problems. Around the same time, the *basis pursuit* (Chen et al., 1998) approach, essentially equivalent to LASSO, was introduced in the signal processing literature, and breakthrough theoretical results of (Candès et al., 2006a) and (Donoho, 2006a) gave rise to the exciting new field of *compressed sensing* that revolutionized signal processing by *exponentially* reducing the number of measurements required for an accurate and computationally efficient recovery of sparse signals, as compared to the standard Shannon-Nyquist theory. In recent years, compressed sensing attracted an enormous amount of interest in signal processing and related communities, and generated a flurry of theoretical results, algorithmic approaches, and novel applications.

In this book, we primarily focus on continuous sparse signals, following the developments in modern sparse statistical modeling and compressed sensing. Clearly, no single book can possibly cover all aspects of these rapidly growing fields. Thus, our goal is to provide a reasonable introduction to the key concepts and survey major recent results in sparse modeling and signal recovery, such as common problem formulations arising in sparse regression, sparse Markov networks and sparse matrix factorization, several basic theoretical aspects of sparse modeling, state-of-the-art algorithmic approaches, as well as some practical applications. We start with an overview of several motivating practical problems that give rise to sparse signal recovery formulations.

1.1 Motivating Examples

1.1.1 Computer Network Diagnosis

One of the central issues in distributed computer systems and networks management is fast, real-time diagnosis of various faults and performance degradations. However, in large-scale systems, monitoring every single component, i.e, every network link, every application, every database transaction, and so on, becomes too costly, or even infeasible. An alternative approach is to collect a relatively small number of overall performance measures using end-to-end transactions, or *probes*, such as *ping* and *traceroute* commands, or end-to-end application-level tests, and then make inferences about the states of individual components. The area of research within the systems management field that focuses on diagnosis of network issues from indirect observations is called *network tomography*, similarly to medical tomography, where health issues are diagnosed based on inferences made from tomographic images of different organs.

In particular, let us consider the problem of identifying network performance bottlenecks, e.g., network links responsible for unusually high end-to-end delays, as discussed, for example, in (Beygelzimer et al., 2007). We assume that $y \in R^m$ is an observed vector of end-to-end transaction delays, $x \in R^n$ is an unobserved vector of link delays, and A is a *routing matrix*, where $a_{ij} = 1$ if the end-to-end test i goes through the link j, and 0 otherwise; the problem is illustrated in Figure 1.2. It is often assumed that the end-to-end delays follow the noisy linear model, i.e.

$$y = Ax + \epsilon, \tag{1.1}$$

where ϵ is the observation noise, that may reflect some other potential causes of end-to-end delays, besides the link delays, as well as possible nonlinear effects. The problem of reconstructing x can be viewed as an *ordinary least squares (OLS)* regression problem, where A is the design matrix and x are the linear regression coefficients found by minimizing the least-squares error, which is also equivalent to maximizing

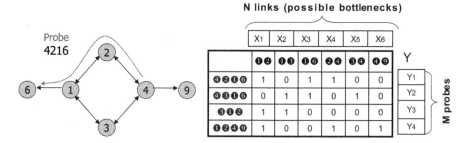

FIGURE 1.2: Example of a sparse signal recovery problem: diagnosing performance bottleneck(s) in a computer network using end-to-end test measurements, or *probes*.

the conditional log-likelihood $\log P(\mathbf{y}|\mathbf{x})$ under the assumption of Gaussian noise ϵ:

$$\min_{\mathbf{x}} \|\mathbf{y} - \mathbf{A}\mathbf{x}\|_2^2.$$

Since the number of tests, m, is typically much smaller than the number of components, n, the problem of reconstructing $\mathbf{x}$ is underdetermined, i.e., there is no unique solution, and thus some regularization constraints need to be added. In case of network performance bottleneck diagnosis, it is reasonable to expect that, at any particular time, there are only a few malfunctioning links responsible for transaction delays, while the remaining links function properly. In other words, we can assume that $\mathbf{x}$ can be well-approximated by a *sparse* vector, where only a few coordinates have relatively large magnitudes, as compared to the rest. Later in this book, we will focus on approaches to enforcing sparsity in the above problem, and discuss sparse solution recovery from a small number of measurements.

1.1.2 Neuroimaging Analysis

We now demonstrate a different kind of application example which arises in medical imaging domain. Specifically, we consider the problem of predicting mental states of a person based on brain imaging data, such as, for example, functional Magnetic Resonance Imaging (fMRI). In the past decade, neuroimaging-based prediction of mental states became an area of active research on the intersection between statistics, machine learning, and neuroscience. A mental state can be cognitive, such as looking at a picture versus reading a sentence (Mitchell et al., 2004), or emotional, such as feeling happy, anxious, or annoyed while playing a virtual-reality videogame (Carroll et al., 2009). Other examples include predicting pain levels experienced by a person (Rish et al., 2010; Cecchi et al., 2012), or learning a classification model that recognizes certain mental disorders such as schizophrenia (Rish et al., 2012a), Alzheimer's disease (Huang et al., 2009), or drug addiction (Honorio et al., 2009).

In a typical "mind reading" fMRI experiment, a subject performs a particular task or is exposed to a certain stimulus, while an MR scanner records the subject's blood-oxygenation-level dependent (BOLD) signals indicative of changes in neural activity, over the entire brain. The resulting full-brain scans over the time period associated with the task or stimulus form a sequence of three-dimensional images, where each image typically has on the order of 10,000-100,000 subvolumes, or *voxels*, and the number of time points, or time repetitions (TRs), is typically on the order of hundreds.

As mentioned above, a typical experimental paradigm aims at understanding changes in a mental state associated with a particular task or a stimulus, and one of the central questions in the modern multivariate fMRI analysis is whether we can predict such mental states given the sequence of brain images. For example, in a recent pain perception study by (Baliki et al., 2009), the subjects were rating their pain level on a continuous scale in response to a quickly changing thermal stimulus applied to their back via a contact probe. In another experiment, associated with the 2007 Pittsburgh Brain Activity Interpretation Competition (Pittsburgh EBC Group, 2007), the task was to predict mental states of a subject during a videogame session, including feeling annoyed or anxious, listening to instructions, looking at a person's face, or performing a certain task within the game.

Given an fMRI data set, i.e. the BOLD signal (voxel activity) time series for all voxels, and the corresponding time series representing the task or stimulus, we can formulate the prediction task as a linear regression problem, where the individual time points will be treated as independent and identically distributed (i.i.d.) samples – a simplifying assumption that is, of course, far from being realistic, and yet often works surprisingly well for predictive purposes. The voxel activity levels correspond to predictors, while the mental state, task, or stimulus is the predicted response variable. More specifically, let $A_1, \ldots, A_n$ denote the set of n predictors, let Y be the response variable, and let m be the number of samples. Then $\mathbf{A} = (\mathbf{a}_1 | \cdots | \mathbf{a}_n)$ corresponds to an $m \times n$ fMRI data matrix, where each $\mathbf{a}_i$ is an m-dimensional vector of the i-th predictor's values, for all m instances, while the m-dimensional vector $\mathbf{y}$ corresponds to the values of the response variable Y, as it is illustrated in Figure 1.3.

As it was already mentioned, in biological applications, including neuroimaging, interpretability of a statistical model is often as important as the model's predictive performance. A common approach to improving a model's interpretability is *variable selection*, i.e. choosing a small subset of predictive variables that are *most relevant* to the response variable. In neuroimaging applications discussed above, one of the key objectives is to discover brain areas that are most relevant to a given task, stimulus, or mental state. Moreover, variable selection, as well as a more general *dimensionality reduction* approach, can significantly improve generalization accuracy of a model by preventing it from overfitting high-dimensional, small-sample data common in fMRI and other biological applications.

A simple approach to variable selection, also known in the machine-learning community as a *filter-based* approach, is to evaluate each predictive variable independently, using some univariate relevance measure, such as, for example, correlation between the variable and the response, or the mutual information between the two. For example, a traditional fMRI analysis approach known as General Linear Models (GLMs) (Friston et al., 1995) can be viewed as filter-based variable selection, since it essentially computes individual correlations between each voxel and

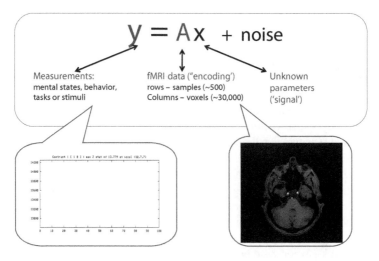

FIGURE 1.3 (See color insert): Mental state prediction from functional MRI data, viewed as a linear regression with simultaneous variable selection. The goal is to find a subset of fMRI voxels, indicating brain areas that are most relevant (e.g., most predictive) about a particular mental state.

the task or stimulus, and then identifies brain areas where these correlations exceed a certain threshold. However, such mass-univariate approach, though very simple, has an obvious drawback, as it completely ignores multivariate interactions, and thus can miss potentially relevant groups of variables that individually do not appear among the top-ranked ones[1]. As it was demonstrated by (Haxby et al., 2001) and others (see, for example, recent work by (Rish et al., 2012b)), highly predictive models of mental states can be built from voxels with sub-maximal activation, that would not be discovered by the traditional GLM analysis. Thus, in recent years, multivariate predictive modeling became a popular alternative to univariate approaches in neuroimaging. Since a combinatorial search over all subsets of voxels in order to evaluate their relevance to the target variable is clearly intractable, a class of techniques, called *embedded methods*, appears to be the best practical alternative to both the univariate selection and the exhaustive search, since it incorporates variable selection into multivariate statistical model learning.

A common example of embedded variable selection is *sparse regression*, where a cardinality constraint restricting the number of nonzero coefficients is added to the original regression problem. Note that in case of linear, or OLS, regression, the resulting sparse regression problem is equivalent to the sparse recovery problem introduced in the network diagnosis example.

---

[1] Perhaps one of the most well-known illustrations of a multi-way interaction among the variables that cannot be detected by looking at any subset of them, not only at the single variables, is the parity check (logical XOR) function over n variables; the parity check response variable is statistically independent of each of its individual inputs, or any subset of them, but is completely determined given the full set of n inputs.

1.1.3   Compressed Sensing

One of the most prominent recent applications of sparsity-related ideas is *compressed sensing*, also known as *compressive sensing*, or *compressive sampling* (Candès et al., 2006a; Donoho, 2006a), an extremely popular and rapidly expanding area of modern signal processing. The key idea behind compressed sensing is that the majority of real-life signals, such as images, audio, or video, can be well approximated by sparse vectors, given some appropriate basis, and that exploiting the sparse signal structure can dramatically reduce the signal acquisition cost; moreover, accurate signal reconstruction can be achieved in a *computationally efficient* way, by using *sparse optimization* methods, discussed later in this book.

Traditional approach to signal acquisition is based on the classical Shannon-Nyquist result stating that in order to preserve information about a signal, one must sample the signal at a rate which is at least twice the signal's *bandwidth*, defined as the highest frequency in the signal's spectrum. Note, however, that such classical scenario gives a worst-case bound, since it does not take advantage of any specific structure that the signal may possess. In practice, sampling at the Nyquist rate usually produces a tremendous number of samples, e.g., in digital and video cameras, and must be followed by a compression step in order to store or transmit this information efficiently. The compression step uses some basis to represent a signal (e.g., Fourier, wavelets, etc.) and essentially throws away a large fraction of coefficients, leaving a relatively few important ones. Thus, a natural question is whether the compression step can be combined with the acquisition step, in order to avoid the collection of an unnecessarily large number of samples.

As it turns out, the above question can be answered positively. Let $s \in R^n$ be a signal that can be represented sparsely in some basis[2] B, i.e. $s = Bx$, where B is an $n \times n$ matrix of basis vectors (columns), and where $x \in R^n$ is a sparse vector of the signal's coordinates with only $k << n$ nonzeros. Though the signal is not observed directly, we can obtain a set of linear measurements:

$$y = Ls = LBx = Ax, \tag{1.2}$$

where L is an $m \times n$ matrix, and $y \in R^m$ is a set of m measurements, or samples, where m can be much smaller than the original dimensionality of the signal, hence the name "compressed sampling". The matrix $A = LB$ is called the *design* or *measurement matrix*. The central problem of compressed sensing is reconstruction of a high-dimensional sparse signal representation x from a low-dimensional linear observation y, as it is illustrated in Figure 1.4a. Note that the problem discussed above describes a *noiseless* signal recovery, while in practical applications there is always some noise in the measurements. Most frequently, Gaussian noise is assumed which leads to the classical linear, or OLS, regression problem, discussed before, though other types of noise are possible. The noisy signal recovery problem is depicted in

---

[2]As mentioned above, Fourier and wavelet bases are two examples commonly used in image processing, though in general finding a good basis that allows for a sparse signal representation is a challenging problem, known as *dictionary learning*, and discussed later in this book.

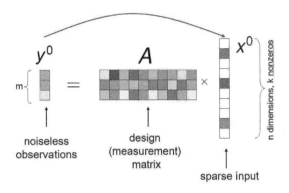

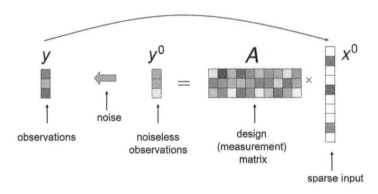

(b) noisy signal recovery

FIGURE 1.4: Compressed sensing – collecting a relatively small number of linear measurements that allow for an accurate reconstruction of a high-dimensional sparse signal: (a) noiseless case, (b) noisy case.

Figure 1.4b, and is equivalent to the diagnosis and sparse regression problems encountered in sections 1.1.1 and 1.1.2, respectively.

1.2 Sparse Recovery in a Nutshell

The following two questions are central to all applications that involve sparse signal recovery: *when* is it possible to recover a high-dimensional sparse signal from a low-dimensional observation vector? And, *how* can we do this in a computationally efficient way? The key results in sparse modeling and compressed sensing identify particular conditions on the design matrix and signal sparsity that allow for an accurate reconstruction of the signal, as well as optimization algorithms that achieve such reconstruction in a computationally efficient way.

Sparse signal recovery can be formulated as finding a minimum-cardinality solution to a constrained optimization problem. In the noiseless case, the constraint is simply $\mathbf{y} = \mathbf{Ax}$, while in the noisy case, assuming Gaussian noise, the solution must satisfy $||\mathbf{y} - \mathbf{y}^*||_2 \leq \epsilon$, where $\mathbf{y}^* = \mathbf{Ax}$ is the (hypothetical) noiseless measurement, and the actual measurement is ϵ-close to it in l_2-norm (Euclidean norm). The objective function is the cardinality of $\mathbf{x}$, i.e. the number of nonzeros, which is often denoted $||\mathbf{x}||_0$ and called l_0-norm of $\mathbf{x}$ (though, strictly speaking, l_0 is not a proper norm), as discussed in the following chapters. Thus, the optimization problems corresponding to noiseless and noisy sparse signal recovery can be written as follows:

$$(\text{noiseless}) \quad \min_{\mathbf{x}} ||\mathbf{x}||_0 \text{ subject to } \mathbf{y} = \mathbf{Ax}, \quad (1.3)$$

$$(\text{noisy}) \quad \min_{\mathbf{x}} ||\mathbf{x}||_0 \text{ subject to } ||\mathbf{y} - \mathbf{Ax}||_2 \leq \epsilon. \quad (1.4)$$

In general, finding a minimum-cardinality solution satisfying linear constraints is an NP-hard combinatorial problem (Natarajan, 1995). Thus, an approximation is necessary to achieve computational efficiency, and it turns out that, under certain conditions, approximate approaches can recover the exact solution.

Perhaps the most widely known and striking result from the compressed sensing literature is that, for a random design matrix, such as, for example, a matrix with i.i.d. Gaussian entries, with high probability, a sparse n-dimensional signal with at most k nonzeros can be reconstructed exactly from only $m = O(k \log(n/k))$ measurements (Candès et al., 2006a; Donoho, 2006a). Thus, the number of samples can be *exponentially smaller* than the signal dimensionality. Moreover, with this number of measurements, a computationally efficient recovery is possible by solving a convex optimization problem:

$$\min_{\mathbf{x}} ||\mathbf{x}||_1 \text{ subject to } \mathbf{y} = \mathbf{Ax}, \quad (1.5)$$

where $||\mathbf{x}||_1 = \sum_{i=1}^{n} |x_i|$ is the l_1-*norm* of $\mathbf{x}$. As shown in chapter 2, the above problem can be reformulated as a linear program and thus easily solved by standard optimization techniques.

More generally, in order to guarantee an accurate recovery, the design matrix does not necessarily have to be random, but needs to satisfy some "nice" properties. The commonly used sufficient condition on the design matrix is the so-called *restricted isometry property (RIP)* (Candès et al., 2006a), which essentially states that a linear transformation defined by the matrix must be almost isometric (recall that an isometric mapping preserves vector length), when restricted to any subset of columns of certain size, proportional to the sparsity k. RIP and other conditions will be discussed in detail in chapter 3.

Furthermore, even if measurements are contaminated by noise, sparse recovery is still *stable* in a sense that recovered signal is a close approximation to the original one, provided that the noise is sufficiently small, and that the design matrix satisfies certain properties such as RIP (Candès et al., 2006a). A sparse signal can be recovered by solving a "noisy" version of the above l_1-norm minimization problem

$$\min_{\mathbf{x}} ||\mathbf{x}||_1 \text{ subject to } ||\mathbf{y} - \mathbf{Ax}||_2 \leq \epsilon. \quad (1.6)$$

The above optimization problem can be also written in two equivalent forms (see, for example, section 3.2 of (Borwein et al., 2006)): either as another constrained optimization problem, for some value of bound t, uniquely defined by ϵ:

$$\min_{\mathbf{x}} ||\mathbf{y} - \mathbf{Ax}||_2^2 \text{ subject to } ||\mathbf{x}||_1 \leq t, \quad (1.7)$$

or as an unconstrained optimization, using the corresponding Lagrangian for some appropriate Lagrange multiplier λ uniquely defined by ϵ, or by t:

$$\min_{\mathbf{x}} \frac{1}{2} ||\mathbf{y} - \mathbf{Ax}||_2^2 + \lambda ||\mathbf{x}||_1. \quad (1.8)$$

In statistical literature, the latter problem is widely known as LASSO regression (Tibshirani, 1996), while in signal processing it is often referred to as *basis pursuit* (Chen et al., 1998).

1.3 Statistical Learning versus Compressed Sensing

Finally, it is important to point out similarities and differences between *statistical* and *engineering* applications of sparse modeling, such as learning sparse models from data versus sparse signal recovery in compressed sensing. Clearly, both statistical and engineering applications involving sparsity give rise to the same optimization problems, that can be solved by the same algorithms, often developed in parallel in both statistical and signal processing communities.

However, statistical learning pursues somewhat different goals than compressed sensing, and often presents additional challenges:

- Unlike compressed sensing, where the measurement matrix can be constructed to have desired properties (e.g., random i.i.d. entries), in statistical learning, the design matrix consists of the observed data, and thus we have little control over its properties. Thus, matrix properties such as RIP are often not satisfied; also note that testing RIP property of a given matrix is NP-hard, and thus computationally infeasible in practice.

- Moreover, when learning sparse models from real-life datasets, it is difficult to evaluate the accuracy of sparse recovery, since the "ground-truth" model is usually not available, unlike in the compressed sensing setting, where the ground truth is the known original signal (e.g., an image taken by a camera).

An easily estimated property of a statistical model is its predictive accuracy on a test data set; however, predictive accuracy is a very different criterion from the *support recovery*, which aims at correct identification of nonzero coordinates in a "ground-truth" sparse vector.

- While theoretical analysis in compressed sensing is often focused on sparse *finite-dimensional* signal recovery and the corresponding conditions on the measurement matrix, the analysis of sparse statistical models is rather focused on *asymptotic consistency* properties, i.e. decrease of some statistical errors of interest with the growing number of dimensions and samples. Three typical performance metrics include: (1) *prediction error* – predictions of the estimated model must converge to the predictions of the true model in some norm, such as l_2-norm; this property is known as model *efficiency*; (2) *parameter estimation error* – estimated parameters must converge to the true parameters, in some norm such as l_2-norm; this property is called *parameter estimation consistency*; and (3) *model-selection error* – the sparsity pattern, i.e. the location of nonzero coefficients, must converge to the one of the true model; this property is also known as *model selection consistency*, or *sparsistency* (also, convergence of the *sign pattern* is called *sign consistency*).

- Finally, recent advances in sparse statistical learning include a wider range of problems beyond the basic sparse linear regression, such as sparse generalized linear models, sparse probabilistic graphical models (e.g., Markov and Bayesian networks), as well as a variety of approaches enforcing more complicated structured sparsity.

1.4   Summary and Bibliographical Notes

In this chapter, we introduced the concepts of sparse modeling and sparse signal recovery, and provided several motivating application examples, ranging from network diagnosis to mental state prediction from fMRI and to compressed sampling of sparse signals. As it was mentioned before, sparse signal recovery dates back to at least 1943, when combinatorial group testing was introduced in the context of Boolean signals and logical-OR measurements (Dorfman, 1943). Recent years have witnessed a rapid growth of the sparse modeling and signal recovery areas, with the particular focus on continuous sparse signals, their linear projections, and l_1-norm regularized reconstruction approaches, triggered by the breakthrough results of (Candès et al., 2006a; Donoho, 2006a) on high-dimensional signal recovery via l_1-based methods, where the number of measurements is logarithmic in the number of dimensions – an exponential reduction when compared to the standard Shannon-Nyquist theory. Efficient l_1-norm based sparse regression, such as LASSO (Tibshirani, 1996) in statistics and its signal-processing equivalent, *basis pursuit* (Chen et al., 1998), are now widely used in various high-dimensional applications.

In the past years, sparsity-related research has expanded significantly beyond the original signal recovery formulation, to include sparse nonlinear regression, such as Generalized Linear Models (GLMs), discussed in chapter 7; sparse probabilistic networks, such as Markov and Bayesian networks, discussed in chapter 8; sparse matrix factorization, such as dictionary learning; sparse PCA and sparse nonnegative matrix factorization (NMF), discussed in chapter 9; and other types of sparse settings.

As it was already mentioned, due to the enormous amount of recent developments in sparse modeling, a number of important topics remain out of scope of this book. One example is the *low-rank matrix completion* – a problem appearing in a variety of applications, including collaborative filtering, metric learning, multi-task learning, and many others. Since the rank minimization problem, similarly to l_0-norm minimization, is intractable, it is common to use its convex relaxation by the *trace norm*, also called the *nuclear norm*, which is the l_1-norm of the vector of singular values. For more details on low-rank matrix learning and trace norm minimization, see, for example, (Fazel et al., 2001; Srebro et al., 2004; Bach, 2008c; Candès and Recht, 2009; Toh and Yun, 2010; Negahban and Wainwright, 2011; Recht et al., 2010; Rohde and Tsybakov, 2011; Mishra et al., 2013) and references therein. Another area we are not discussing here in detail is *sparse Bayesian learning* (Tipping, 2001; Wipf and Rao, 2004; Ishwaran and Rao, 2005; Ji et al., 2008), where alternative priors, beyond the Laplacian (equivalent to the l_1-norm regularizer), are introduced in order to enforce the solution sparsity. Also, besides several applications of sparse modeling that we will discuss herein, there are multiple others that we will not be able to include, in the fields of astronomy, physics, geophysics, speech processing, and robotics, just to name a few.

For further references on recent developments in the field, as well as for tutorials and application examples, we refer the reader to the online repository available at the Rice University website[3], and to other online resources[4]. Several recent books focus on particular aspects of sparsity; for example, (Elad, 2010) provides a good introduction to sparse representations and sparse signal recovery, with a particular focus on image-processing applications. A classical textbook on statistical learning by (Hastie et al., 2009) includes, among many other topics, introduction to sparse regression and its applications. Also, a recent book by (Bühlmann and van de Geer, 2011) focuses specifically on sparse approaches in high-dimensional statistics. Moreover, various topics related to compressed sensing are covered in several recently published monographs and edited collections (Eldar and Kutyniok, 2012; Foucart and Rauhut, 2013; Patel and Chellappa, 2013).

---

[3]http://dsp.rice.edu/cs.
[4]See, for example, the following blog at http://nuit-blanche.blogspot.com.

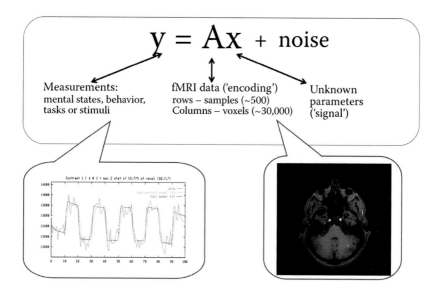

FIGURE 1.3: Mental state prediction from functional MRI data, viewed as a linear regression with simultaneous variable selection. The goal is to find a subset of fMRI voxels, indicating brain areas that are most relevant (e.g., most predictive) about a particular mental state.

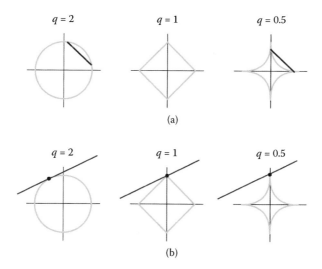

FIGURE 2.3: (a) Level sets $||\mathbf{x}||_q^q = 1$ for several values of q. (b) Optimization of (P_q) as inflation of the origin-centered l_q-balls until they meet the set of feasible points $\mathbf{A}\mathbf{x} = \mathbf{y}$.

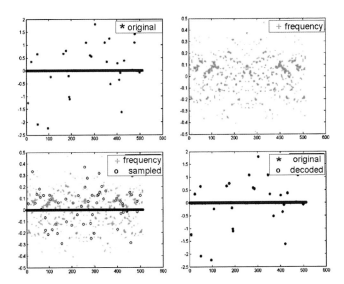

FIGURE 3.2: A one-dimensional example demonstrating perfect signal reconstruction based on l_1-norm. Top-left (a): the original signal $\mathbf{x}_0$; top-right (b): (real part of) the DFT of the original signal, $\hat{\mathbf{x}}_0$; bottom-left (c): observed spectrum of the signal (the set of Fourier coefficients); bottom-right (d): solution to P_1': exact recovery of the original signal.

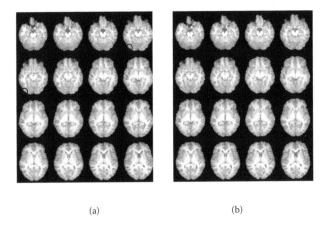

(a)   (b)

FIGURE 6.3: Brain maps showing absolute values of the Elastic Net solution (i.e. coefficients x_i of the linear model) for the Instruction target variable in PBAIC dataset, for subject 1 (radiological view). The number of nonzeros (active variables) is fixed to 1000. The two panels show the EN solutions (maps) for (a) $\lambda_2 = 0.1$ and (b) $\lambda_2 = 2$. The clusters of nonzero voxels are bigger for bigger λ_2, and include many, but not all, of the $\lambda_2 = 0.1$ clusters. Note that the highlighted (red circle) cluster in (a) is identified by EN with $\lambda_2 = 0.1$, but not in the $\lambda_2 = 2.0$ model.

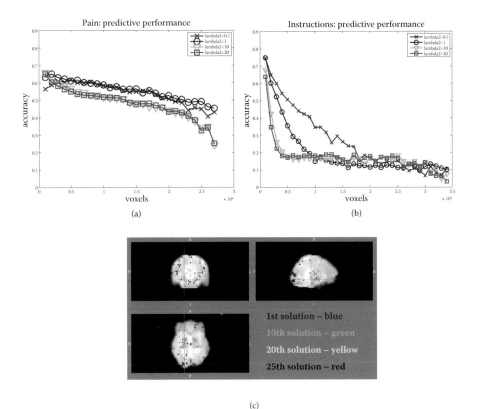

(a)

(b)

(c)

FIGURE 6.5: Predictive accuracy of the subsequent "restricted" Elastic Net solutions, for (a) pain perception and (b) "Instructions" task in PBAIC. Note very slow accuracy degradation in the case of pain prediction, even for solutions found after removing a significant amount of predictive voxels, which suggests that pain-related information is highly distributed in the brain (also, see the spatial visualization of some solutions in Figure (c)). The opposite behavior is observed in the case of the "Instruction" — a sharp decline in the accuracy after a few first "restricted" solutions are deleted, and very localized predictive solutions shown earlier in Figure 6.3.

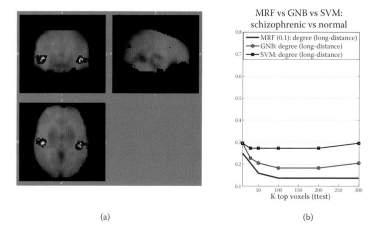

(a)   (b)

FIGURE 8.2: (a) FDR-corrected 2-sample t-test results for (normalized) degree maps, where the null hypothesis at each voxel assumes no difference between the schizophrenic vs. normal groups. Red/yellow denotes the areas of low p-values passing FDR correction at $\alpha = 0.05$ level (i.e., 5% false-positive rate). Note that the mean (normalized) degree at those voxels was always (significantly) *higher* for normals than for schizophrenics. (b) Gaussian MRF classifier predicts schizophrenia with 86% accuracy using just 100 top-ranked (most-discriminative) features, such as voxel degrees in a functional network.

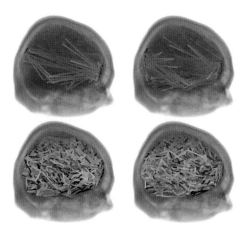

FIGURE 8.3: Structures learned for cocaine addicted (left) and control subjects (right), for sparse Markov network learning method with variable-selection via $\ell_{1.2}$ method (top), and without variable-selection, i.e., standard graphical lasso approach (bottom). Positive interactions are shown in blue, negative interactions are shown in red. Notice that structures on top are much sparser (density 0.0016) than the ones on the bottom (density 0.023) where the number of edges in a complete graph is $\approx 378,000$.

Chapter 2

Sparse Recovery: Problem Formulations

The focus of this chapter is on optimization problems that arise in sparse signal recovery. We start with a simple case of noiseless linear measurements, which is later extended to more realistic noisy recovery formulation(s). Since the ultimate problem of finding the sparsest solution – the solution with the smallest number of nonzeros, also called the l_0-norm – is computationally hard (specifically, NP-hard) due to its nonconvex combinatorial nature, one must resort to approximations. Two main approximation approaches are typically used in sparse recovery: the first one is to address the original NP-hard combinatorial problem via approximate methods, such as greedy search, while the second is to replace the intractable problem with its convex relaxation that is easy to solve. In other words, one can either solve the exact problem approximately, or solve an approximate problem exactly. In this chapter, we primarily discuss the second approach – convex relaxations–while the approximate methods such as greedy search are discussed later in chapter 5. We consider the family of l_q-norm bounds on the l_0-norm, and focus on the l_1-norm, in particular, since it is the only norm in the l_q-family that is both sparsity-inducing and convex. Finally, we discuss the Bayesian (point estimation) perspective on sparse signal recovery and sparse statistical learning, which yields the *maximum a posteriory (MAP)* parameter estimation. The MAP approach gives rise to regularized optimization, where the negative log-likelihood and the prior on the parameters (i.e., on the signal we wish to recover) correspond to a loss function and a regularizer, respectively.

2.1   Noiseless Sparse Recovery

We follow the notation introduced before: $\mathbf{x} = (x_1, \ldots, x_n) \in \mathbb{R}^n$ is an un-observed sparse signal, $\mathbf{y} = (y_1, \ldots, y_m) \in \mathbb{R}^m$ is a vector of measurements, or observations, and $\mathbf{A} = \{a_{ij}\} \in \mathbb{R}^{m \times n}$ is a design matrix. Also, $\mathbf{A}_{i,:}$ and $\mathbf{A}_{:,j}$ will denote the i-th row and the j-th column of $\mathbf{A}$, respectively. However, when there is no ambiguity, and the notation is clearly defined in a particular context, we will often use $\mathbf{a}_i$ as a shorthand for the i-th row or the i-th column-vector of the matrix $\mathbf{A}$. In general, boldface upper-case letters, such as $\mathbf{A}$, will denote matrices, boldface lower-case letters, such as $\mathbf{x}$, $\mathbf{y}$, and $\mathbf{a}_i$, will denote vectors, and regular (non-bold) lower-case letters, such as x_i, will denote scalars. For example, note the difference between a scalar x_i, usually denoting the i-th coordinate of the vector $\mathbf{x}$, and the vector $\mathbf{x}_i$ denoting the i-th vector in some set of vectors.

The simplest problem setting we are going to start with is the noiseless signal recovery from a set of linear measurements, i.e. solving for $\mathbf{x}$ the system of linear equations:

$$\mathbf{Ax} = \mathbf{y}. \tag{2.1}$$

It is usually assumed that $\mathbf{A}$ is a full-rank matrix, and thus for any $\mathbf{y} \in \mathbb{R}^m$, the above system of linear equations has a solution. Note that when the number of unknown variables, i.e. dimensionality of the signal, exceeds the number of observations, i.e. when $n \geq m$, the above system is underdetermined, and can have infinitely many solutions. In order to recover the signal $\mathbf{x}$, it is necessary to further constrain, or *regularize*, the problem. This is usually done by introducing an objective function, or regularizer $R(\mathbf{x})$, that encodes additional properties of the signal, with lower values corresponding to more desirable solutions. Signal recovery is then formulated as a constrained optimization problem:

$$\min_{\mathbf{x} \in \mathbb{R}^n} R(\mathbf{x}) \text{ subject to } \mathbf{y} = \mathbf{Ax}. \tag{2.2}$$

For example, when the desired quality is sparsity, $R(\mathbf{x})$ can be defined as the number of nonzero elements, or the cardinality of $\mathbf{x}$, also called the l_0-*norm*, and denoted $||\mathbf{x}||_0$. Note, however, that the l_0-norm is not a proper norm, formally speaking, as we discuss shortly. The rationale for calling the cardinality of a vector its l_0-norm is explained below.

In general, l_q-norms for particular values of q, denoted $||\mathbf{x}||_q$, or, more precisely, their q-th power $||\mathbf{x}||_q^q$, are frequently used as regularizers. We will now take a closer look at l_q-norms and their properties (also, see Appendix). Recall that for a $q \geq 1$, the l_q-norm, also called just q-norm of a vector $\mathbf{x} \in \mathbb{R}^n$, is defined as follows:

$$||\mathbf{x}||_q = \left(\sum_{i=1}^{n} |x_i|^q \right)^{1/q}. \tag{2.3}$$

Among most frequently used l_q-norms are the l_2-norm (Euclidean norm)

$$||\mathbf{x}||_2 = \sqrt{\sum_{i=1}^{n} |x_i|^2} \tag{2.4}$$

and the l_1-norm

$$||\mathbf{x}||_1 = \sum_{i=1}^{n} |x_i|. \tag{2.5}$$

As it is shown in Appendix, section A.1, for $q \geq 1$, the functions defined in eq. 2.3 are indeed proper norms, i.e., they satisfy the standard norm properties. When $0 < q < 1$, the function defined in eq. 2.3 is not a proper norm since it violates the triangle inequality (again, see section A.1 in Appendix), although, for convenience sake, it is still quite often called l_q-norm in the literature, ignoring the abuse of terminology.

We now go back to the cardinality of a vector and its relation to $||l_q||$-norms. The function $||\mathbf{x}||_0$ referred to as l_0-norm of $\mathbf{x}$ is defined as a limit of $||\mathbf{x}||_q^q$, i.e. l_q-norms to the q-th power, when $q \to 0$:

$$||\mathbf{x}||_0 = \lim_{q \to 0} ||\mathbf{x}||_q^q = \lim_{q \to 0} \sum_{i=1}^{p} |x_i|^q = \sum_{i=1}^{p} \lim_{q \to 0} |x_i|^q. \tag{2.6}$$

For each x_i, when $q \to 0$, $|x_i|^q \to I(x_i)$, the indicator function, which is 0 at $\mathbf{x} = 0$ and 1 otherwise. Figure 2.1 illustrates this convergence, showing how $|x_i|^q$ for several decreasing values of q gets closer and closer to the indicator function.

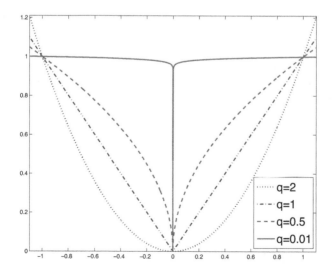

FIGURE 2.1: $||\mathbf{x}||_0$-norm as a limit of $||\mathbf{x}||_q^q$ when $q \to 0$.

Thus, $||\mathbf{x}||_0 = \sum_{i=1}^{p} I(x_i)$, which gives exactly the number of nonzero elements of vector $\mathbf{x}$, or its *cardinality*[1]. Using the cardinality function, we can now write the problem of sparse signal recovery from noiseless linear measurements as follows:

$$(P_0): \quad \min_{\mathbf{x}} ||\mathbf{x}||_0 \ \text{ subject to } \ \mathbf{y} = \mathbf{A}\mathbf{x} \qquad (2.7)$$

2.2   Approximations

As it was mentioned before, the problem (P_0) defined in the eq. 2.7 is NP-hard (Natarajan, 1995), i.e. no known algorithm can solve it efficiently, in polynomial time. Therefore, approximations are necessary. The good news is that, under appropriate conditions, the optimal, or close to the optimal, solution(s) can still be recovered efficiently by certain approximate techniques.

The following two types of approximate approaches are commonly used. The first one is to apply a *heuristic-based search*, such as a greedy search, in order to explore the solution space of (P_0). For example, one can start with a zero vector and keep adding nonzero coordinates one by one, selecting at each step the coordinate that leads to the best improvement in the objective function (i.e., greedy coordinate descent). In general, such heuristic search methods are not guaranteed to find the global optimum. However, in practice, they are simple to implement, very efficient computationally and often find sufficiently good solutions. Moreover, under certain conditions, they are even guaranteed to recover the optimal solution. Greedy approaches to the sparse signal recovery problem will be discussed later in this book.

An alternative approximation technique is the *relaxation approach* based on replacing an intractable objective function or constraint by a tractable one. For example, *convex relaxations* approximates a non-convex optimization problem by a convex one, i.e. by a problem with convex objective and convex constraints. Such problems are known to be "easy", i.e. there exists a variety of efficient optimization methods for solving convex problems. Clearly, besides being easy to solve, e.g., convex, the relaxed version of the (P_0) problem must also enforce solution sparsity. In the following sections, we discuss l_q-norm based relaxations, and show that the l_1-norm occupies a unique position among them, combining convexity with sparsity.

---

[1] Note again, that the name l_0-norm can be somewhat misleading since $||x||_0$ is clearly not a proper norm, as it violates the absolute homogeneity property: indeed, $||\alpha x||_0 = ||x||_0$ for $\alpha \neq 0$; in other words, l_0-"norm" is not scale-sensitive.

2.3 Convexity: Brief Review

We will now briefly review the notion of convexity, before starting the discussion of convex relaxations for our sparse recovery problem (P_0).

Given two vectors, $\mathbf{x}_1 \in \mathbb{R}^n$ and $\mathbf{x}_2 \in \mathbb{R}^n$, and a scalar $\alpha \in [0,1]$, vector $\mathbf{x} = \alpha\mathbf{x}_1 + (1-\alpha)\mathbf{x}_2$ is called a *convex combination* of $\mathbf{x}_1$ and $\mathbf{x}_2$. A set S is called a *convex set* if any convex combination of its elements belongs to the set, i.e.

$$\forall \mathbf{x}_1, \mathbf{x}_2 \in S, \ \ \forall \alpha \in [0,1], \ \ \mathbf{x} \in S \ if \ \mathbf{x} = \alpha\mathbf{x}_1 + (1-\alpha)\mathbf{x}_2.$$

In other words, a set is convex if for any two points in the set, the line segment connecting them also lies within the set.

A function $f(\mathbf{x}) : S \to \mathbb{R}$ defined on a convex set S in a vector space is called a *convex function* if

$$\forall \mathbf{x}_1, \mathbf{x}_2 \in S \ and \ \forall \alpha \in [0,1], f(\alpha\mathbf{x}_1 + (1-\alpha)\mathbf{x}_2) \leq \alpha f(\mathbf{x}_1) + (1-\alpha)f(\mathbf{x}_2).$$

In other words, the line segment connecting a pair of points in the plot of a convex function is always above the plot of that function (see Figure 2.2). Another way to interpret this geometrically is to say that the set of points above the function's plot (also called the *epigraph* of a function) is convex. A function is called *strictly convex* if the above inequality is strict, i.e.

$$f(\alpha\mathbf{x}_1 + (1-\alpha)\mathbf{x}_2) < \alpha f(\mathbf{x}_1) + (1-\alpha)f(\mathbf{x}_2),$$

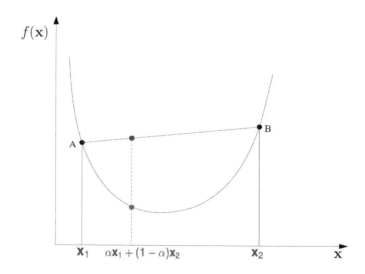

FIGURE 2.2: A convex function.

assuming $\mathbf{x}_1 \neq \mathbf{x}_2$ and $0 < \alpha < 1$. An important property of a convex function is that any of its local minima is also a global one. Moreover, a strictly convex function has a unique global minimum.

A *convex optimization problem* is minimization of a convex function over a convex set of feasible solutions defined by the constraints. Due to the properties of the convex objective functions, convex problems are easier to solve than general optimization problems. Convex optimization is a traditional area of research in optimization literature, with many efficient solution techniques developed in the past years.

2.4   Relaxations of (P_0) Problem

We will now return to our problem of main interest, (P_0), i.e. cardinality minimization with linear constraints. It is easy to see that the constraint $\mathbf{y} = \mathbf{Ax}$ yields a convex feasible set; indeed, given two feasible solutions $\mathbf{x}_1$ and $\mathbf{x}_2$, satisfying this constraint, any convex combination of them is also feasible, since

$$\mathbf{A}(\alpha\mathbf{x}_1 + (1-\alpha)\mathbf{x}_2) = \alpha\mathbf{Ax}_1 + (1-\alpha)\mathbf{Ax}_2 = \alpha\mathbf{y} + (1-\alpha)\mathbf{y} = \mathbf{y}.$$

Thus, in order to relax (P_0) to a convex problem, we only need to replace the objective function, $||\mathbf{x}||_0$, by a convex one.

Herein, we will focus on the l_q-norms as possible relaxations of l_0. More precisely, we will consider the q-th power of l_q-norms, i.e. the functions $||\mathbf{x}||_q^q$, as regularization functions $R(\mathbf{x})$ in the general problem setting in eq. 2.2. As Figure 2.1 shows for the one-dimensional case, these functions are convex for $q \geq 1$, and nonconvex for $q < 1$. For example, l_2-norm, or Euclidean norm (defined in eq. 2.4), is perhaps one of the most well-known and commonly used l_q-norms, and is a natural first choice as a relaxation of l_0-norm. We get

$$(P_2): \quad \min_{\mathbf{x}} ||\mathbf{x}||_2^2 \text{ subject to } \mathbf{y} = \mathbf{Ax} \tag{2.8}$$

There are several advantages to using this objective: the function $||\mathbf{x}||_2^2$ is strictly convex and thus always has a unique minimum. Moreover, the solution to the problem (P_2) is available in a closed form. Indeed, to solve the problem (P_2), we write its *Lagrangian*:

$$\mathcal{L}(\mathbf{x}) = ||\mathbf{x}||_2^2 + \lambda^T(\mathbf{y} - \mathbf{Ax})$$

where λ is an m-dimensional vector with all coordinates equal to λ, and the optimality conditions

$$\frac{\partial \mathcal{L}(\mathbf{x})}{\partial \mathbf{x}} = 2\mathbf{x} + \mathbf{A}^T\lambda = 0.$$

This gives the unique optimal solution $\mathbf{x}^* = -\frac{1}{2}\mathbf{A}^T\lambda$. Since $\mathbf{x}^*$ must satisfy $\mathbf{y} = \mathbf{Ax}^*$, we get $\lambda = -2(\mathbf{AA}^T)^{-1}\mathbf{y}$, and thus

$$\mathbf{x}^* = -\frac{1}{2}\mathbf{A}^T\lambda = \mathbf{A}^T(\mathbf{AA}^T)^{-1}\mathbf{y}.$$

This closed form solution to the (P_2) problem is also known as a *pseudo-inverse solution* of $\mathbf{y} = \mathbf{Ax}$ when $\mathbf{A}$ has more columns than rows (as mentioned earlier, it is also assumed that $\mathbf{A}$ is full-rank, i.e. all of its rows are linearly independent). However, despite its convenient properties, $||\mathbf{x}||_2^2$ objective has a serious drawback: it turns out that the optimal solution (P_2) is practically never sparse, and thus cannot be used as a good approximation in sparse signal recovery.

2.5   The Effect of l_q-Regularizer on Solution Sparsity

To understand why the l_2-norm does not promote the solution sparsity while the l_1-norm does, and to understand the convexity and sparsity-inducing properties of l_q-norms in general, let us consider the geometry of a (P_q) problem, where $||\mathbf{x}||_q^q$ replaces the original cardinality objective $||\mathbf{x}||_0$:

$$(P_q): \quad \min_{\mathbf{x}} ||\mathbf{x}||_q^q \text{ subject to } \mathbf{y} = \mathbf{Ax}. \qquad (2.9)$$

Sets of vectors with same value of the function $f(\mathbf{x})$, i.e. $f(\mathbf{x}) = const$, are called the *level sets* of $f(\mathbf{x})$. For example, the level sets of $||\mathbf{x}||_q^q$ function are vector sets with same l_q-norm. Figure 2.3a shows examples of the level sets $||\mathbf{x}||_q^q = 1$ for several values of q. A set of vectors satisfying $||\mathbf{x}||_q^q \leq r^q$ is called an l_q-ball of radius r; its "surface" (set boundary) is the corresponding level set $||\mathbf{x}||_q^q = r^q$. Note that the corresponding l_q-balls bounded by the level sets in Figure 2.3a are *convex* for $q \geq 1$ (line segments between a pair of its points belong to the ball), and *nonconvex* for $0 < q < 1$ (line segments between a pair of its points do not always belong to the ball).

From a geometric point of view, solving the optimization problem (P_q) is equivalent to "blowing up" l_q-balls with the center at the origin, i.e., increasing their radius, starting from 0, until they touch the hyperplane $\mathbf{Ax} = \mathbf{y}$, as it is shown in Figure 2.3b. The resulting point is the minimum l_q-norm vector that is also a feasible point, i.e. it is the optimal solution of (P_q).

Note that when $q \leq 1$, l_q-balls have sharp "corners" on the coordinate axis, corresponding to sparse vectors, since some of their coordinates are zero, but l_q-balls for $q > 1$ do not have this property. Thus, for $q \leq 1$, l_q-balls are likely to meet the hyperplane $\mathbf{Ax} = \mathbf{y}$ at the corners, thus producing sparse solutions, while for $q > 1$ the intersection practically never occurs at the axes, and thus solutions are not sparse. Note that this is an intuitive geometric illustration of the l_q-norm properties, rather than a formal argument; a more formal analysis is provided later in this book.

In summary, we need to approximate the intractable combinatorial l_0-norm objective by a function that would be easier to optimize, but that would also produce sparse solutions. Within the family of $||\mathbf{x}||_q^q$ functions, only those with $q \geq 1$ are convex, but only those with $0 < q \leq 1$ are sparsity-enforcing. The only function within that family that has both useful properties is therefore $||\mathbf{x}||_1$, i.e. the l_1-norm.

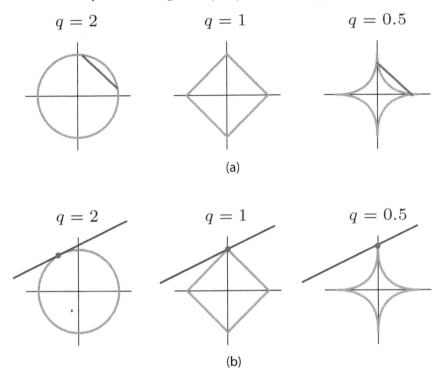

(a)

(b)

FIGURE 2.3 (See color insert): (a) Level sets $||\mathbf{x}||_q^q = 1$ for several values of q. (b) Optimization of (P_q) as inflation of the origin-centered l_q-balls until they meet the set of feasible points $\mathbf{Ax} = \mathbf{y}$.

This unique combination of sparsity and convexity is the reason for the widespread use of l_1-norms in the modern sparse signal recovery field. In the noiseless case, the l_1-norm relaxation of intractable (P_0) problem, stated as the (P_1) problem below, became the main focus of theoretical and algorithmic studies:

$$(P_1): \quad \min_{\mathbf{x}} ||\mathbf{x}||_1 \text{ subject to } \mathbf{y} = \mathbf{Ax}. \qquad (2.10)$$

2.6 l_1-norm Minimization as Linear Programming

Note that the problem (P_1) can be also reformulated as a linear program (Chen et al., 1998), a very well studied class of optimization problems with efficient solution techniques. Indeed, let us introduce new nonnegative variables $\mathbf{u}, \mathbf{v} \in \mathbb{R}^n$, such that $\mathbf{x} = \mathbf{u} - \mathbf{v}$, and u_i are nonzero only at positive entries of $\mathbf{x}$, while v_i are nonzero only at negative entries of $\mathbf{x}$. Using the notation $\mathbf{z} = [\mathbf{u}^T, \mathbf{v}^T]^T \in \mathbb{R}^{2n}$, we now get $||\mathbf{x}||_1 = \sum_i^{2n} z_i$, and $\mathbf{Ax} = \mathbf{A}(\mathbf{u} - \mathbf{v}) = [\mathbf{A}, -\mathbf{A}]\mathbf{z}$. Then (P_1) is equivalent to the following Linear Program (LP):

$$\min_{\mathbf{z}} \sum_i^{2n} z_i \text{ subject to } \mathbf{y} = [\mathbf{A}, -\mathbf{A}]\mathbf{z} \text{ and } \mathbf{z} \geq 0. \quad (2.11)$$

Now we just need to verify that, for an optimal solution, the above assumption about $\mathbf{u}$ and $\mathbf{v}$ having non-overlapping supports indeed holds, with $\mathbf{u}$ and $\mathbf{v}$ corresponding to positive-only and negative-only entries in $\mathbf{x}$, respectively. Indeed, let us assume the opposite, then there exist u_j and v_j, for some j, that are both nonzero; also, due to the nonnegativity constraint above, $u_j > 0$ and $v_j > 0$. Without loss of generality, we can assume that $u_j \geq v_j$, and replace u_j by $u'_j = u_j - v_j$ and v_j by $v'_j = 0$ in the solution to the above LP. Clearly, the nonnegativity constraint is still satisfied, as well as the linear constraint $\mathbf{y} = [\mathbf{A}, -\mathbf{A}]$, since $A_j u_j - A_j v_j = A_j u'_j - A_j v'_j$, so the new solution is still feasible. However, it also reduces the objective function by $2v_j$, which contradicts the optimality of the initial solution. Thus, we showed that $\mathbf{u}$ and $\mathbf{v}$ are indeed non-overlapping, i.e. the initial assumption about decomposing $\mathbf{x}$ into positive-only and negative-only parts holds, and thus the problem (P_1) is indeed equivalent to the above linear program in eq. 2.11.

2.7 Noisy Sparse Recovery

So far we considered an idealistic setting with noiseless measurements. However, in practical applications, such as image processing or statistical data modeling, the measurement noise is unavoidable. Thus, the linear equation constraint $\mathbf{Ax} = \mathbf{y}$ must be relaxed in order to allow for some discrepancy between the "ideal" observation $\mathbf{Ax}$ and its realistic noisy version. For example, it is typically replaced by the inequality $||\mathbf{y} - \mathbf{Ax}||_2 \leq \epsilon$, stating that the actual measurement vector $\mathbf{y}$ is ϵ-close, in Euclidean norm, to the (unavailable) noiseless measurement $\mathbf{Ax}$. (From a probabilistic perspective, Euclidean norm, as we discuss in more detail later, arises from the Gaussian noise assumption about observations; other noise models lead to a wider class of distances.) Such relaxation is also helpful for exploring approximate solutions to the original noiseless problem (P_0). Moreover, a relaxation is necessary in cases when there are more observations than unknowns, i.e. more rows than columns in $\mathbf{A}$, a common situation in the classical regression setting, and thus the linear system $\mathbf{Ax} = \mathbf{y}$ may have no solutions. The noisy sparse recovery problem can be then written as follows:

$$(P_0^\epsilon): \quad \min_{\mathbf{x}} ||\mathbf{x}||_0 \text{ subject to } ||\mathbf{y} - \mathbf{Ax}||_2 \leq \epsilon. \quad (2.12)$$

The corresponding l_1-norm relaxation of the l_0-norm objective, similarly to the noiseless case, can be written as

$$(P_1^\epsilon): \quad \min_{\mathbf{x}} ||\mathbf{x}||_1 \text{ subject to } ||\mathbf{y} - \mathbf{A}\mathbf{x}||_2 \leq \epsilon. \qquad (2.13)$$

Note that, equivalently, we could impose the constraint on the square of the l_2-norm, rather than the l_2-norm, i.e. $||\mathbf{y} - \mathbf{A}\mathbf{x}||_2^2 \leq \nu$, where $\nu = \epsilon^2$. Then, using an appropriate Lagrange multiplier $\lambda(\epsilon)$, denoted simply λ below, we can also rewrite the above problem as an unconstrained minimization:

$$(P_1^\lambda): \quad \min_{\mathbf{x}} \frac{1}{2}||\mathbf{y} - \mathbf{A}\mathbf{x}||_2^2 + \lambda||\mathbf{x}||_1, \qquad (2.14)$$

or, for an appropriate parameter $t(\epsilon)$, denoted simply as t, the same problem can be also rewritten as follows:

$$(P_1^t): \quad \min_{\mathbf{x}} ||\mathbf{y} - \mathbf{A}\mathbf{x}||_2^2 \text{ subject to } ||\mathbf{x}||_1 \leq t. \qquad (2.15)$$

As we already mentioned in the previous chapter, the above l_1-norm regularized problem, especially in its two latter forms, (P_1^λ) and (P_1^t), is widely known as the *LASSO (Least Absolute Shrinkage and Selection Operator)* (Tibshirani, 1996) in statistical literature, and as the *Basis Pursuit* (Chen et al., 1998) in signal processing. Note that, similarly to noiseless recovery problem in the previous section, the problem (P_1^λ) can be reformulated as a *quadratic programming (QP)*, which can be solved by standard optimization toolboxes:

$$\min_{\mathbf{x}_+, \mathbf{x}_- \in \mathbb{R}_+^n} \frac{1}{2}||\mathbf{y} - \mathbf{A}\mathbf{x}_+ + \mathbf{A}\mathbf{x}_-||_2^2 + \lambda(1^T\mathbf{x}_+ + 1^T\mathbf{x}_-). \qquad (2.16)$$

The geometry of the LASSO problem is demonstrated in Figure 2.4, for the two special cases: (a) $n \leq m$, the low-dimensional case, when the number of measurements exceeds the number of variables, and (b) the high-dimensional case, i.e. $n > m$. In both cases, the l_1-norm constraint is shown as the diamond-shaped area with "pointy edges" corresponding to sparse feasible solutions, and the level sets of the quadratic objective function in eq. 2.15 have different shape depending on whether the number of variables exceeds the number of observations or not.

In the low-dimensional case when $n \leq m$, the quadratic objective function in eq. 2.15 has the unique minimum $\hat{\mathbf{x}} = (\mathbf{A}^T\mathbf{A})^{-1}\mathbf{A}^T\mathbf{y}$, provided that the matrix $\mathbf{A}$ has a full *column-rank*, i.e. its *columns* are linearly independent. This can be easily verified by taking the derivative of the objective function in eq. 2.15:

$$f(\mathbf{x}) = ||\mathbf{y} - \mathbf{A}\mathbf{x}||_2^2 = (\mathbf{y} - \mathbf{A}\mathbf{x})^T(\mathbf{y} - \mathbf{A}\mathbf{x}),$$

and setting it to zero

$$\frac{\partial f(\mathbf{x})}{\partial \mathbf{x}} = -2\mathbf{A}^T(\mathbf{y} - \mathbf{A}\mathbf{x}) = 0.$$

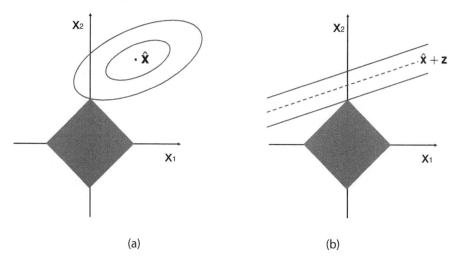

FIGURE 2.4: Geometry of the noisy sparse recovery via l_1-minimization, i.e. problem (P_1^ϵ), in two cases: (a) $n \leq m$, or a low-dimensional case, as in classical regression setting with more observations than unknowns; here $\hat{\mathbf{x}} = (\mathbf{A}^T\mathbf{A})^{-1}\mathbf{A}^T\mathbf{y}$ is the unique ordinary least squares (OLS) solution; (b) $n > m$, or high-dimensional case, with more unknowns than observations; in this case, there are multiple solutions of the form $\hat{\mathbf{x}} + \mathbf{z}$, $\forall \mathbf{z} \in N(\mathbf{A})$, where $N(\mathbf{A})$ is the null space of $\mathbf{A}$.

This gives us the unique solution $\hat{\mathbf{x}} = (\mathbf{A}^T\mathbf{A})^{-1}\mathbf{A}^T\mathbf{y}$, also called the Ordinary Least Squares (OLS) solution, since minimization of the above objective is equivalent to solving the basic OLS regression problem:

$$(OLS): \quad \min_{\mathbf{x}} ||\mathbf{y} - \mathbf{A}\mathbf{x}||_2^2.$$

The OLS solution is also known as a pseudo-inverse solution of $\mathbf{y} = \mathbf{A}\mathbf{x}$ when $\mathbf{A}$ has more rows than columns[2]. Note that the OLS solution to the least-squares problem above exists even if the solution to the system of linear equations $\mathbf{y} = \mathbf{A}\mathbf{x}$ does not, which can happen in cases when there are more equations than unknowns, and thus least squares relaxation is used in such cases. The level sets of the objective function $||\mathbf{y} - \mathbf{A}\mathbf{x}||_2^2 = const$ start with the single point $\hat{\mathbf{x}}$ at the minimum, and for larger values of the function correspond to ellipses, as shown in Figure 2.4a.

On the other hand, when $m < n$, and $\mathbf{A}$ is full row-rank, linear system $\mathbf{y} = \mathbf{A}\mathbf{x}$ always has a solution, such as the minimum-l_2-norm solution $\hat{\mathbf{x}} = \mathbf{A}^T(\mathbf{A}\mathbf{A}^T)^{-1}\mathbf{y}$ of (P_2), i.e. pseudo-inverse of $\mathbf{y} = \mathbf{A}\mathbf{x}$ in cases of more columns than rows. Moreover, there are infinitely many solutions of the form $\hat{\mathbf{x}} + \mathbf{z}$, for all $\mathbf{z} \in N(\mathbf{A})$, where $N(\mathbf{A})$ is the null space of $\mathbf{A}$, i.e. a set of all points such that $\mathbf{A}\mathbf{z} = 0$, and thus

---

[2]Note that the solution of the (P_2) problem mentioned above was also called a pseudo-inverse solution, but for the case when $\mathbf{A}$ had more columns than rows, and was full *row-rank*, i.e. all its *rows* were linearly independent.

$\mathbf{A}(\hat{\mathbf{x}} + \mathbf{z}) = \mathbf{A}\hat{\mathbf{x}} = \mathbf{y}$. All these solutions form a hyperplane $\mathbf{A}\mathbf{x} = \mathbf{y}$ (line in two-dimensions, as shown in Figure 2.4b), corresponding to the minimal-value level set of the objective function, i.e. $||\mathbf{y} - \mathbf{A}\mathbf{x}||_2^2 = 0$. Any other level set $||\mathbf{y} - \mathbf{A}\mathbf{x}||_2^2 = const$ will corresponds to two hyperplanes parallel to $\mathbf{A}\mathbf{x} = \mathbf{y}$, at the same fixed distance from it (see Figure 2.4b).

Let $t_0 = \min_{\mathbf{z} \in N(\mathbf{A})} ||\hat{\mathbf{x}} + \mathbf{z}||_1$ be the minimum l_1-norm achieved by the solutions to the linear system $\mathbf{y} = \mathbf{A}\mathbf{x}$ (or the single solution $\hat{\mathbf{x}}$ in case of $m \geq n$). We will assume that $t < t_0$ in the eq. 2.15, otherwise the l_1-norm constraint is vacuous, i.e. the LASSO problem becomes equivalent to the unconstrained ordinary least squares. Then the least-squares solution(s) are located outside the feasible region, and thus any solution $\mathbf{x}^*$ to eq. 2.15 must be on the boundary of that region, i.e. when the level sets of the objective function first meet the feasible region, which implies $||\mathbf{x}^*||_1 = t$. Note that the diamond-shaped feasible region $||x||_1 < t$ tends to meet the level sets of the quadratic objective function at the diamond's vertices (this is most easily seen in two-dimensional cases, but also generalizes to multidimensional cases), which correspond to sparse solutions. This example illustrates the intuition behind the sparsity-enforcing property of the l_1-norm constraint, similarly to the noiseless sparse recovery problem.

Note that the feasible region defined by the constraint $||\mathbf{x}||_1 \leq t$ is convex; since the quadratic objective function is also convex, the solution set of the LASSO problem is convex as well, i.e. any convex combination of the solution is also a solution. The properties of LASSO solutions discussed above can be now summarized as follows:

Theorem 2.1. *(Osborne et al., 2000b) 1. If $m \geq n$ (more samples than unknowns), then the LASSO problem in eq. 2.15 has a unique solution $\mathbf{x}^*$, and $||\mathbf{x}^*||_1 = t$.*
2. If $n > m$ (less samples than unknowns), then a solution $\mathbf{x}^*$ of the LASSO problem exists and $||\mathbf{x}^*||_1 = t$ for any solution. 3. Also, if $\mathbf{x}_1^*$ and $\mathbf{x}_2^*$ are both solutions of the LASSO problem, then their convex combination $\alpha\mathbf{x}_1^* + (1 - \alpha)\mathbf{x}_2^*$, where $0 \leq \alpha \leq 1$, is also a solution.

We now mention a couple of other basic properties of LASSO solutions, such as *optimality conditions* and *solution (or regularization) path*. The optimality conditions, obtained by setting to zero the subgradient[3] of the LASSO objective, are commonly used in the literature; for example, see (Fuchs, 2005), as well as more recent work by (Mairal and Yu, 2012), among several others.

Lemma 2.2. *(Optimality conditions) A vector $\hat{\mathbf{x}} \in R^n$ is a solution to the LASSO problem (P_1^λ) given in eq. 2.14 if and only if the following conditions hold for all $i \in \{1, \ldots, n\}$:*

$$\mathbf{a}_i^T(\mathbf{y} - \mathbf{A}\hat{\mathbf{x}}) = \lambda \, sign(\hat{\mathbf{x}}_i) \quad if \ \hat{\mathbf{x}}_i \neq 0,$$

$$|\mathbf{a}_i^T(\mathbf{y} - \mathbf{A}\hat{\mathbf{x}})| \leq \lambda \quad if \ \hat{\mathbf{x}}_i = 0,$$

where $\mathbf{a}_i$ is the i-th column of the matrix $\mathbf{A}$, and where $sign(x) = 1$ if $x > 0$, $sign(x) = -1$ if $x < 0$, and $sign(x) = 0$ otherwise.

---

[3]Chapter 5 discusses subgradients in more detail.

A *regularization path*, or *solution path*, of LASSO is the sequence of all solutions obtained for varying regularization parameter $\lambda > 0$:

$$\hat{\mathbf{x}}(\lambda) = \arg\min_{\mathbf{x}} f(\mathbf{x}, \lambda) = \arg\min_{\mathbf{x}} \frac{1}{2}||\mathbf{y} - \mathbf{A}\mathbf{x}||_2^2 + \lambda||\mathbf{x}||_1.$$

An important property of the LASSO solution path is its *piecewise linearity*, which allows for efficient active-set methods such as the homotopy algorithm of (Osborne et al., 2000a) and LARS (Efron et al., 2004), presented later in chapter 5. An intuition behind the piecewise linearity of the LASSO path is as follows. Let $\lambda_1 < \lambda_2$ be two sufficiently close values of λ, so that going from the solution $\hat{\mathbf{x}}(\lambda_1)$ to $\hat{\mathbf{x}}(\lambda_2)$ does not (yet) require any coordinate of $\hat{\mathbf{x}}$ to change its sign. Then it is easy to see that, for all $0 \leq \alpha \leq 1$, the optimality conditions stated above are satisfied for $\lambda = \alpha\lambda_1 + (1 - \alpha)\lambda_2$ and for $\hat{\mathbf{x}} = \alpha\hat{\mathbf{x}}(\lambda_1) + (1 - \alpha)\hat{\mathbf{x}}(\lambda_2)$, and thus $\hat{\mathbf{x}}(\lambda) = \hat{\mathbf{x}}$. In other words, the regularization path between λ_1 and λ_2 is a line, as long as there is no sign change in the corresponding solutions.

2.8 A Statistical View of Sparse Recovery

We now discuss an alternative view at the sparse recovery problem, which typically arises in the context of sparse statistical modeling. As it was already mentioned in chapter 1, in statistical learning settings, the columns of the design matrix $\mathbf{A}$ correspond to random variables A_j, called predictors, and the rows of $\mathbf{A}$ correspond to samples, i.e. to the observations of predictive variables, which are often assumed to be independent and identically distributed, or i.i.d. Also, the entries of $\mathbf{y}$ correspond to observations of another random variable, Y, called the response variable. The task is to learn a statistical model capable of predicting the response, given the predictors. For example, in a functional MRI analysis (fMRI), columns of $\mathbf{A}$ typically correspond to BOLD signals at particular voxels in 3D brain images, the rows correspond to particular time points at which the subsequent 3D images are obtained, and the entries of $\mathbf{y}$ correspond to a stimulus, such as, for example, the temperature in thermal pain studies, measured at the same time points. The unobserved vector $\mathbf{x}$ represents the parameters of a statistical model describing the relationship between the response and predictors, such as, for example, a linear model with Gaussian noise, or OLS regression.

We now define a general statistical learning framework, as follows. Let $\mathbf{Z} = (\mathbf{A}, \mathbf{y})$ denote the observed data, i.e. a collection of m samples containing the values of the n predictors and the corresponding response, and let $M(\mathbf{x})$ denote a model with parameters $\mathbf{x}$. The standard model selection approach assumes a *loss function* $L_M(\mathbf{Z}, \mathbf{x})$, or simply $L(\mathbf{Z}, \mathbf{x})$, describing the discrepancy between the observed data and their approximation provided by the model, e.g. sum-squared loss between the linear model estimate $\hat{\mathbf{y}} = \mathbf{A}\mathbf{x}$ and the actual observations $\mathbf{y}$. The model selection is commonly viewed as minimization of the loss function with respect to the parameters **x** in order to find the model that best fits the data. However, when the number of parameters n exceeds the number of samples m, such approach is prone to *overfitting* the data, i.e. learning a model that can represent the training data really well, but will fail to generalize to the test data, i.e. previously unseen data that are presumably coming from the same data distribution. Since the ultimate objective of statistical learning is indeed a good generalization accuracy of a model, an additional *regularization* constraint is typically added to the above optimization problem in order to prevent overfitting by restricting the parameter space when searching for the minimum-loss solution. The model selection problem can be generally stated as follows, where regularization function is denoted as $R(\mathbf{x})$:

$$\min_{\mathbf{x}} L(\mathbf{Z}, \mathbf{x}) \text{ subject to } R(\mathbf{x}) \leq t, \qquad (2.17)$$

which can also be rewritten in two equivalent formulations, where ϵ and λ are uniquely determined by t, and vice versa:

$$\min_{\mathbf{x}} R(\mathbf{x}) \text{ subject to } L(\mathbf{Z}, \mathbf{x}) \leq \epsilon, \qquad (2.18)$$

or, using an appropriate Lagrange multiplier λ,

$$\min_{\mathbf{x}} L(\mathbf{Z}, \mathbf{x}) + \lambda R(\mathbf{x}). \qquad (2.19)$$

We now illustrate a probabilistic interpretation of both loss and regularization functions. Namely, we assume that the model $M(\mathbf{x})$ describes the probability distribution $P(\mathbf{Z}|\mathbf{x})$ of the data, where $\mathbf{x}$ are the parameters of this distribution. Also, using the Bayesian approach, we assume a prior distribution of the parameters $P(\mathbf{x}|\lambda)$, with a *hyper-parameter* λ, which is assumed to be fixed for now. The model learning problem is then commonly stated as the *maximum a posteriori (MAP)* parameter estimation, i.e. finding a vector of parameters $\mathbf{x}$ maximizing the joint probability $P(\mathbf{Z}, \mathbf{x}) = P(\mathbf{Z}|\mathbf{x})P(\mathbf{x}|\lambda)$, or, equivalently, minimizing the negative log-likelihood

$$\min_{\mathbf{x}} -\log[P(\mathbf{Z}|\mathbf{x})P(\mathbf{x}|\lambda)],$$

which can be written as

$$\min_{\mathbf{x}} -\log P(Z|\mathbf{x}) - \log P(\mathbf{x}|\lambda). \qquad (2.20)$$

Note that the MAP formulation of the learning problem gives rise to the regularized loss minimization problem described above, with the loss function $L(\mathbf{Z}, \mathbf{x}) = -\log P(\mathbf{Z}|\mathbf{x})$, which is lower for the models that have higher likelihood, i.e. fit the data better, and the regularization function $R(\mathbf{x}, \lambda) = -\log P(\mathbf{x}|\lambda)$ determined by the prior on the model parameters.

The MAP approach to learning statistical models from data generalizes a wide range of problem formulations. For example, the noisy sparse recovery problem (P_1), i.e. l_1-regularized sum-squared loss minimization, also called sparse linear regression, can be viewed as a particular instance of this approach, with *linear Gaussian observations* and *Laplace prior* on the parameters. Namely, let us assume that elements of $\mathbf{y}$ are i.i.d. random variables following the Gaussian (normal) distribution

$$N_{\mu,\sigma}(z) = \frac{1}{\sqrt{2\pi}\sigma} e^{-\frac{1}{2}(z-\mu)^2},$$

with the standard deviation $\sigma = 1$ and the mean $\mu = \mathbf{a}_i\mathbf{x}$, where $\mathbf{a}_i$ is the i-th row in the matrix $\mathbf{A}$, i.e.

$$P(y_i|\mathbf{a}_i\mathbf{x}) = \frac{1}{\sqrt{2\pi}} e^{-\frac{1}{2}(y_i-\mathbf{a}_i\mathbf{x})^2}.$$

Then the likelihood of the data $\mathbf{Z} = (\mathbf{A}, \mathbf{y})$, assuming the above model with the fixed parameters $\mathbf{x}$, is $P(\mathbf{A}, \mathbf{y}|\mathbf{x}) = P(\mathbf{y}|\mathbf{A}\mathbf{x})P(\mathbf{A})$. The negative log-likelihood loss function can be thus written as

$$L(\mathbf{y}, \mathbf{A}, \mathbf{x}) = -\log P(\mathbf{y}|\mathbf{A}\mathbf{x}) - \log P(\mathbf{A}) = -\log \prod_{i=1}^{m} P(y_i|\mathbf{a}_i\mathbf{x}) - \log P(\mathbf{A}) =$$

$$= \frac{1}{2}\sum_{i=1}^{m}(y_i - \mathbf{a}_i\mathbf{x})^2 + const,$$

where $const = \log \sqrt{2\pi} - \log P(\mathbf{A})$ does not depend on $\mathbf{x}$, and thus can be ignored in the objective function in eq. 2.19. Thus, our linear Gaussian assumption about $\mathbf{y}$ leads to the sum-squared loss function

$$L(\mathbf{y}, \mathbf{A}, \mathbf{x}) = \frac{1}{2}\sum_{i=1}^{m}(y_i - \mathbf{a}_i\mathbf{x})^2 = \frac{1}{2}||y - \mathbf{A}\mathbf{x}||_2^2.$$

Next, let us assume that parameters x_i, $i = 1, \ldots, n$ are i.i.d. random variables following the *Laplace prior* with the hyperparameter λ:

$$p(z) = \frac{\lambda}{2} e^{-\lambda|z|}.$$

Examples of Laplace distribution for different values of λ are shown in Figure 2.5. Note that, as λ increases, more probability weight is assigned to values closer to zero. With the Laplace priors on the parameters, the regularizer can be written as

$$R(\mathbf{x}, \lambda) = -\log P(\mathbf{x}|\lambda) = -\log \prod_{i}^{n} P(x_i|\lambda) =$$

$$= \lambda \sum_{i}^{n} |x_i| + const,$$

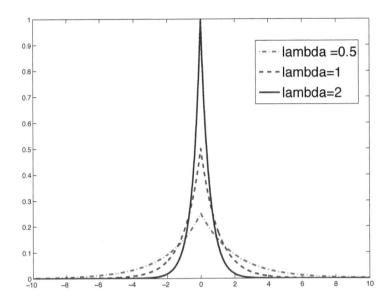

FIGURE 2.5: Laplace distribution for different values of the λ parameter.

where $const = -\log \frac{\lambda}{2}$ does not depend on $\mathbf{x}$ and can be ignored in our regularized loss minimization problem in eq. 2.19, leading to the familiar l_1-norm regularizer

$$R(\mathbf{x}, \lambda) = \lambda \sum_{i}^{n} |x_i| = ||\mathbf{x}||_1.$$

Thus, as we mentioned above, the l_1-norm regularized linear regression problem can be derived as a MAP estimation in a linear Gaussian observation model with Laplace priors on the parameters.

2.9 Beyond LASSO: Other Loss Functions and Regularizers

Sparse linear regression was just one example illustrating the regularized log-likelihood maximization approach. Here we provide a very brief overview of several other types of log-likelihood losses and regularizers that often appear in the literature, and are summarized in Figure 2.6. Note, however, that *the examples below are only used as a quick preview of the material covered later in this book*, and, at this point, the reader is not expected to immediately develop a deep understanding of all new concepts involved in those examples from this very "compressed" summary.

Sparse GLM regression (chapter 7). One natural way to extend the classical linear regression is to go beyond the standard Gaussian noise assumption to a general class

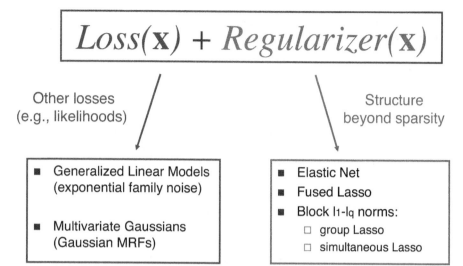

FIGURE 2.6: Several examples of regularized optimization problems, with different loss and regularization functions.

of *exponential-family* noise distributions, which includes, besides Gaussian, a wide variety of distributions, such as logistic, multinomial, exponential, and Poisson, just to name a few. When the observations are discrete-valued, e.g., binary or categorical, take only positive values, or otherwise are not well-described by the Gaussian model, exponential-family distributions may be more appropriate. We will discuss recovery of sparse signals with exponential-family observations in chapter 7. For now, we only would like to mention that recovering a signal $\mathbf{x}$ from linear observations $\mathbf{Ax}$ contaminated by an exponential-family noise yields the so-called *Generalized Linear Model (GLM)* regression problems. In such models, the log-likelihood loss is

$$L = -\log \sum_{i=1}^{n} P(y_i|\Theta_i) = \sum_{i=1}^{n} B(y_i, \mu_i),$$

where $\Theta_i = \mathbf{a}_i \mathbf{x}$ is the *natural parameter* of a particular exponential-family distribution ($\mathbf{a}_i$ denotes here the i-th row of $\mathbf{A}$), $\mu_i(\Theta_i)$ is the corresponding mean parameter, and $B(y_i, \mu_i)$ denotes the so-called *Bregman divergence* between the observation y_i and its mean parameter μ_i. Bregman divergences generalize the Euclidean distance associated with Gaussian noise, and each type of exponential-family noise is associated with its own Bregman divergence.

Sparse GMRFs (chapter 8). Another commonly used type of data distribution is *multivariate Gaussian*, since its log-likelihood is involved in learning sparse Gaussian Markov Networks, also called *sparse Gaussian Markov Random Fields (GMRFs)*. This area received much attention in the recent machine-learning and statistical

literature. The sparse GMRF learning problem is discussed in full detail in chapter
8; here we just mention that the corresponding negative log-likelihood loss function
is given by

$$L = -\log \sum_{i=1}^{n} P(Z^i|C) = tr(SC) - \log\det(C),$$

where S is the empirical covariance matrix, and C is the inverse covariance (or precision) matrix. It is usually assumed that the data are centered to have zero mean,
and thus the covariance (or inverse covariance) parameters are sufficient to define the
multivariate Gaussian distribution.

Beyond l_1-regularization (chapter 6). Finally, there are various types of regularizers, beyond the l_1-norm, that are used in different situations. As it was already mentioned, the l_1-norm regularizer arises as the negative log-likelihood of the Laplace
prior. Similarly, the squared l_2-norm regularizer $||\mathbf{x}||_2^2$, just like the sum-squared loss
function discussed before, arises from the negative log-likelihood of the Gaussian
prior, and is used in *ridge regression*:

$$\min_{\mathbf{x}} \frac{1}{2}||\mathbf{y} - \mathbf{A}\mathbf{x}||_2^2 + \lambda||\mathbf{x}||_2^2. \tag{2.21}$$

More generally, the q-th power of the l_q-norm, $||\mathbf{x}||_q^q$, for $q \geq 1$, is a family of
regularizers used in the *bridge regression* (Frank and Friedman, 1993; Fu, 1998)
that includes both LASSO and ridge, and corresponds to the prior $P_{\lambda,q}(\mathbf{x}) \sim
C(\lambda, q)e^{-\lambda||\mathbf{x}||_q^q}$:

$$\min_{\mathbf{x}} \frac{1}{2}||\mathbf{y} - \mathbf{A}\mathbf{x}||_2^2 + \lambda||\mathbf{x}||_q^q. \tag{2.22}$$

Finally, several other regularizers extending the basic l_1-norm penalty were introduced recently in order to model more sophisticated, or *structured* sparsity patterns.
For example, sometimes we may want to select not the individual variables, but
groups (subsets) of them, assuming a given set of groups to select from. Group sparsity is enforced via so-called *block-penalties* such as l_1/l_2 or l_1/l_∞, discussed in
chapter 6. Yet another example is the Elastic Net penalty, which is a convex combination of the l_1- and the l_2-norm. When the groups of variables to be selected together
are not given in advance, however, some variables tend to be highly correlated; we
may want to include or exclude such correlated variables together, as a group. The
Elastic Net penalty combines the sparsity-enforcing property of the l_1-norm with
the *grouping effect* added by the l_2-norm, which enforces similar-magnitude regression coefficients on correlated variables. This property is important for interpret-
ing sparse models in biological and neuroimaging applications, just to name a few,
where predictor variables tend to be highly correlated, and the objective is to identify whole groups of such relevant variables. The Elastic Net is also discussed in
chapter 6.

2.10   Summary and Bibliographical Notes

In this chapter, we discussed optimization problems that formalize the sparse signal recovery from linear measurements, for both noiseless and noisy settings. The ultimate sparse recovery problem is defined as finding the sparsest vector satisfying a set of linear constraints, or, in other words, minimizing the l_0-norm of such vector, which is known to be an NP-hard combinatorial problem (Natarajan, 1995; see also Appendix). In order to make the problem tractable, various convex relaxations can be used, such as l_q-norms with $q \geq 1$; however, among all such relaxations, only the l_1-norm is sparsity-promoting, which explains its popularity in the sparse modeling and signal recovery fields. Signal recovery from *noisy* measurements (Candès et al., 2006b; Donoho, 2006b), in particular, is most relevant in practical applications, such as image processing, sensor networks, biology, and medical imaging, just to name a few; see (Resources, 2010) for a comprehensive list of references on compressed sensing and its recent applications. Similarly to the noiseless case, this problem can be formulated as the l_0-norm minimization, subject to linear inequality constraints (instead of linear equations), which allows for including noise in the observations. Again, the l_1-norm relaxation is commonly used instead of the l_0-norm in order to make the problem tractable while still preserving sparsity of solution(s).

Regularized loss minimization problems discussed herein can be also interpreted as maximum a posteriory (MAP) probability parameter estimation in a model with an appropriate probability of observed data and an appropriate prior on the parameters. For example, the noisy sparse signal recovery via l_1-regularized sum-squared loss minimization, known as the LASSO (Tibshirani, 1996) or the Basis Pursuit (Chen et al., 1998), can be viewed as a MAP estimation problem where linear observations are disturbed by the Gaussian noise, and where the Laplace prior is assumed on the parameters. Other types of data distributions and priors give rise to a wide variety of problems, as it was outlined in section 2.9, such as: structured sparsity, which involves regularizers beyond the l_1-norm, such as group LASSO, simultaneous LASSO, fused LASSO, and the Elastic Net, discussed in more detail in chapter 6; sparse Generalized Linear Models and sparse Gaussian Markov Networks, which replace the objective function by the negative log-likelihood functions for the corresponding statistical models, as discussed in chapters 7 and 8, respectively.

Moreover, various other loss functions have been also explored in the literature; though we are not going to discuss all of them in this book, a brief review is provided below. For example, a well-known alternative to the LASSO is the *Dantzig selector* proposed by (Candès and Tao, 2007), which replaces the LASSO's loss function in eq. 2.14 by $||\mathbf{A}^T(\mathbf{y} - \mathbf{Ax})||_\infty$, i.e. by the maximum absolute value of the inner product of the current residual $(\mathbf{y} - \mathbf{Ax})$ with all the predictors. Properties of the Dantzig selector, as well as its relation to LASSO, have been extensively analyzed in the recent literature; see, for example, (Meinshausen et al., 2007; Efron et al., 2007; Bickel, 2007; Cai and Lv, 2007; Ritov, 2007; Friedlander and Saunders, 2007; Bickel et al., 2009; James et al., 2009; Koltchinskii, 2009; Asif and Romberg, 2010). There are also several other popular loss functions, such as, for example, *Huber loss* (Huber, 1964), which combines quadratic and linear pieces, and produces robust regression models; and *hinge loss* $L(\mathbf{y}, \mathbf{Ax}) = \sum_{i=1}^{m} \max(0, 1 - y\mathbf{a}_i\mathbf{x})$, where $\mathbf{a}_i$ denotes the i-th row of $\mathbf{A}$; and also squared hinge loss discussed in (Rosset and Zhu, 2007), frequently used in classification problems where each output $\mathbf{y}_i$ is either 1 or -1.

Also, note that the collection of different *structured-sparsity* promoting regularizers considered in chapter 6 is not exhaustive. For example, a prominent regularizer that remains out of scope of this book is the *nuclear norm*, also known as the *trace norm*; it is used in multiple applications, e.g., in recent work on multivariate regression (Yuan et al., 2007) and clustering (Jalali et al., 2011), among many other examples; see also the bibliography section of the previous chapter for more references on the trace norm.

We also would like to mention another important group of techniques that extend the basic LASSO approach, namely, the methods concerned with improving the *asymptotic consistency* properties of LASSO. This topic has generated a considerable amount of interest in sparse modeling literature; see (Knight and Fu, 2000; Greenshtein and Ritov, 2004; Donoho, 2006c,b; Meinshausen, 2007; Meinshausen and Bühlmann, 2006; Zhao and Yu, 2006; Bunea et al., 2007; Wainwright, 2009), and other references below. The issue with the basic LASSO is that, in general, it is not guaranteed to be consistent in terms of model selection – e.g., it may select "extra" variables that do not belong to the true model (Lv and Fan, 2009); also, due to parameter shrinkage, LASSO may produce biased parameter estimates. Herein, we briefly mention some popular methods for improving the asymptotic consistency of the LASSO. For example, the *relaxed LASSO* (Meinshausen, 2007) is a two-stage procedure that first chooses a subset of variables via the LASSO, and then applies the LASSO again on this subset, now with less "competition" among the variables and thus with a smaller parameter λ selected via cross-validation; this leads to less shrinkage, and, as a result, to less biased parameter estimates. An alternative approach, known as *smoothly clipped absolute deviation (SCAD)* (Fan and Li, 2005), modifies the LASSO penalty in order to reduce the shrinkage of large coefficients; however, it is non-convex. Yet another convex approach, is the *adaptive LASSO* of (Zou, 2006) that uses adaptive weights to penalize different coefficients in the l_1-norm penalty. Moreover, methods such as the *bootstrap Lasso* (Bach, 2008a) and *stability-selection* (Meinshausen and Bühlmann, 2010) use the bootstrap approach, i.e. learn multiple LASSO models on subsets of data, and then include in the model only the intersection of nonzero coefficients (Bach, 2008a), or sufficiently frequently selected nonzero coefficients (Meinshausen and Bühlmann, 2010). This approach eliminates "unstable" coefficients and improves model-selection consistency of LASSO, as well as stability of solutions to the choice of the λ parameter. A recent book by (Bühlmann and van de Geer, 2011) provides a comprehensive treatment of these approaches.

Chapter 3

Theoretical Results (Deterministic Part)

This chapter provides an overview of several theoretical results that are central to the sparse signal recovery. As already mentioned, the key questions in this field are: What types of signals can be reconstructed accurately from an incomplete set of observations? What conditions on the design matrix and on the signal would guarantee an accurate reconstruction? We will start this chapter with a brief overview of the seminal results by (Donoho, 2006a) and (Candès et al., 2006a) that address these questions, and provide a couple of illustrative examples. Note that there exist much earlier theoretical results on sparse signal recovery, dating back to 1989 – see, for example, (Donoho and Stark, 1989; Donoho and Huo, 2001). However, the recent work by (Donoho, 2006a; Candès et al., 2006a) achieved a significant improvement over these earlier results, reducing the number of samples required for the exact sparse signal recovery from square root to logarithmic in the signal's dimension, by going from deterministic to probabilistic setting.

We will next discuss when the solutions of the l_0- and l_1-norm minimization problems, (P_0) and (P_1), are unique, and what are sufficient conditions for these two problem to be equivalent, i.e. when the exact l_0-norm recovery can be achieved by its l_1-norm relaxation. More specifically, we will focus on the following properties of the design matrix: spark, mutual coherence, null-space property, and the restricted isometry property (RIP). We establish the deterministic part of the classical result stating that RIP guarantees accurate recovery of a sparse signal in both noiseless and noisy scenarios.

3.1   The Sampling Theorem

The classical result in signal processing that specifies general conditions for an accurate signal recovery is the well-known Nyquist-Shannon *sampling theorem*, stating that *a perfect signal recovery from discrete samples is possible when the sampling frequency is greater than twice the signal bandwidth (the maximal frequency contained by the signal)*. This important result was derived independently by multiple researchers, including the work of (Nyquist, 1928; Shannon, 1949), as well as (Kotelnikov, 1933; Whittaker, 1915, 1929), and several others. The theorem is sometimes called the Whittaker-Nyquist-Kotelnikov-Shannon sampling theorem, or just "the sampling theorem".

More specifically, a signal is given by some function $f(t)$ over a continuous domain, such as time or space, and sampling refers to the process of converting said signal to a discrete sequence of numbers, by taking measurements at particular points in time or space. For example, let $f(t)$ be a signal in a time domain, and let B hertz be the highest frequency (bandwidth) the signal contains; then, according to the sampling theorem, sampling $f(t)$ at a series of points spaced $\frac{1}{2B}$ seconds apart allows to reconstruct the signal exactly. The discrete version of this theorem (see Appendix) is applicable in the case of a (generally, complex-valued) discrete input signal $\mathbf{x} \in \mathbb{C}^N$, such as, for example, an image given by a finite set of N pixels. The theorem states that, in order to reconstruct a discrete signal, the number of acquired Fourier samples must match the size of the signal N.

However, in many practical applications, the number of samples dictated by this theory may be quite high, making signal acquisition too costly and also requiring compression for transmission and storage of the collected samples. Note, however, that the conditions of the theorem are sufficient rather than necessary, and thus a perfect signal recovery might be still possible with a lower number of samples, provided that the input signals have some special properties, such as low "effective" dimensionality. Studying such signals and an associated reduction in the number of samples needed for signal reconstruction is the main focus of the recently developed compressed sensing field.

3.2   Surprising Empirical Results

In 2006, Candes, Romberg, and Tao published a paper presenting the following puzzling empirical phenomenon that may seem to contradict the conventional wisdom stated by the sampling theorem (Candès et al., 2006a). They experimented with a simulated image called the *Shepp-Logan phantom* (Figure 3.1a); the image was created by (Shepp and Logan, 1974) as a standard for computerized tomography (CT) image reconstruction simulations of the human head. The goal is to

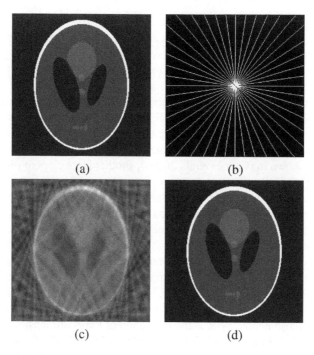

FIGURE 3.1: (a) The Shepp-Logan phantom test image. (b) Sampling domain in the frequency plane; Fourier coefficients are sampled along 22 radial lines. (c) Reconstruction obtained by setting small Fourier coefficients to zero (minimum-energy reconstruction). (d) Exact reconstruction by minimizing the total variation.

reconstruct this two-dimensional image from samples of its discrete Fourier transform (see Appendix) on a star-shaped domain (Figure 3.1b), typically used in practice where imaging devices collect samples along radial lines at a few angles. In this example, 512 samples are gathered along each of 22 radial lines.

A common approach used in medical imaging is to set the Fourier coefficients at unobserved frequencies to zero, and to apply next the inverse Fourier transform; this is the so-called "minimum-energy" reconstruction. An image reconstructed by such an approach is shown in Figure 3.1c. It is clearly of low quality and contains multiple artifacts, due to severe undersampling of the original image. Note, however, that the original image in Figure 3.1a appears to have a certain structure, i.e., it is mostly piecewise constant, or smooth – it does not have too many abrupt changes in the pixel intensity values. As shown in (Candès et al., 2006a), this image can actually be reconstructed *perfectly* from the given samples (see Figure 3.1d), if instead of using the standard minimum-energy approach, we minimize the following convex function known as *total variation*, which measures smoothness of an image:

$$\|g\|_{TV} = \sum_{t_1, t_2} \sqrt{|D_1 g(t_1, t_2)|^2 + |D_2 g(t_1, t_2)|^2}, \tag{3.1}$$

where $g(t_1, t_2), 0 \leq t_1, t_2 \leq N - 1$ is the observed discrete function (e.g., an image) over $N \times N$ points specified by (t_1, t_2), and $D_1 g$, $D_2 g$ are finite differences between the function values at neighboring points, specified as $D_1 g = g(t_1, t_2) - g(t_1 - 1, t_2)$ and $D_2 g = g(t_1, t_2) - g(t_1, t_2 - 1)$. The real surprise is that this perfect reconstruction was achieved by using a much smaller number of samples (almost 50 times smaller, as discussed in (Candès et al., 2006a)) than the sampling theorem would require!

The Shepp-Logan reconstruction phenomenon prompted further theoretical investigation of the conditions on the signals and samples that would allow a much lower sampling rate than the one dictated by the Nyquist-Shannon theorem. As it was further shown in (Candès et al., 2006a), another large class of discrete-time signals $\mathbf{x} \in \mathbb{C}^N$ that allow perfect recovery from only partial knowledge of their Fourier coefficients are *sparse* signals, where only k out of N values are nonzero. As in the example above, a perfect reconstruction is achieved from a sub-Nyquist sampling rate, as long as it exceeds some lower bound, by minimizing another convex function – in this case, the l_1-norm of the signal.

We now present a simple simulation example (only a few lines of MATLAB code), illustrating l_1-norm based signal recovery. Let us first introduce some notation. Let $\mathbb{Z}_N = \{1, \cdots, N\}^1$, and let $\mathbf{x} \in \mathbb{R}^N$. We denote the set of all coordinates with nonzero values by $K = \{i \in \mathbb{Z}_N | \mathbf{x}_i \neq 0\}$, and the size of this set by $|K| = \#\{i \in K\}$. We will also denote by $\mathbf{x}|_K$ the restriction of the N-dimensional vector $\mathbf{x}$ on a subset $K \subset \mathbb{Z}_N$. In our simulation, $N = 512$ and $|K| = 30$. We randomly choose $k = |K|$ numbers from $\mathbb{Z}_N$ to be the support of the vector signal $\mathbf{x}$, and assign k randomly selected real values to the corresponding coordinates of $\mathbf{x}_0$. We now have the "ground-truth" signal that needs to be recovered later (see Figure 3.2a). We then apply the Discrete Fourier Transform, or DFT (see Appendix, section A.2),[2] to $\mathbf{x}_0$, and obtain $\hat{\mathbf{x}}_0 = \mathcal{F}(\mathbf{x}_0)$ (Figure 3.2b). Given the N-dimensional DFT vector $\hat{\mathbf{x}}_0$, let us choose a subset S of 60 coordinates, and restrict $\hat{\mathbf{x}}_0$ on S, setting all other coordinates to zero (see Figure 3.2c). This is the observable spectrum of the signal (the set of Fourier coefficients). We will now attempt to reconstruct the signal from its partially observed Fourier spectrum by solving the following optimization problem:

$$(P_1') : \quad \min_{\mathbf{x}} ||\mathbf{x}||_1 \text{ subject to } \mathcal{F}(\mathbf{x})|_S = \hat{\mathbf{x}}_0|_S. \tag{3.2}$$

The solution to the above problem is shown in Figure 3.2d; as we can see, the original signal is recovered exactly from the incomplete set of Fourier coefficients. In fact, as our simulations show, the exact recovery happens with overwhelming probability over the choice of a particular subset S. This is yet another surprising example that appears to contrast the (discrete version of) Whittaker-Nyquist-Kotelnikov-Shannon sampling theorem (see Appendix, theorem A.1).

---

[1]Note that $\mathbb{Z}_N$ is usually defined as $\{0, \cdots, N - 1\}$ since it denotes the finite field of order N, but we will slightly abuse this notation here and use with base 1 instead of 0.

[2]Strictly speaking, we only use the real part of DFT, i.e., the so-called Discrete Cosine Transform, or DCT.

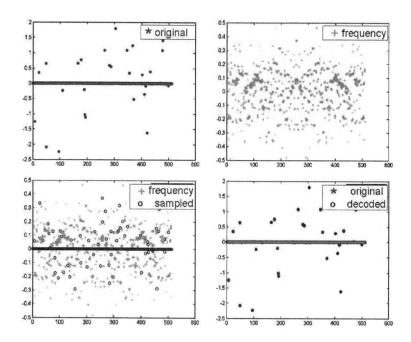

FIGURE 3.2 (See color insert): A one-dimensional example demonstrating perfect signal reconstruction based on l_1-norm. Top-left (a): the original signal x_0; top-right (b): (real part of) the DFT of the original signal, $\hat{x}_0$; bottom-left (c): observed spectrum of the signal (the set of Fourier coefficients); bottom-right (d): solution to P_1': exact recovery of the original signal.

3.3 Signal Recovery from Incomplete Frequency Information

The surprising empirical phenomenon described above, namely, the fact that an exact signal recovery is possible from a number of samples below the bound specified by the standard sampling theorem, was theoretically justified in seminal papers by (Candès et al., 2006a) and (Donoho, 2006a). Thus, we will call this phenomenon the *Donoho-Candès-Romberg-Tao (DCRT) phenomenon*.

As was shown in the above papers, the two key factors contributing to the DCRT phenomenon are: *sparsity* (or, more generally, some low-dimensional structure) of the original signal x, and *randomness* of the observed subset of frequencies in $\hat{x}$.

Definition 1. *An N-dimensional vector $x \in \mathbb{C}^N$ is called k-sparse if it has at most k nonzero coordinates. The set of nonzero coordinates of x is called its* support set, *denoted $supp(x)$. Given an arbitrary (not necessarily sparse) vector x, we obtain its k-sparse approximation by keeping k coordinates that have the largest absolute values, and setting the rest of the coordinates to zero.*

The main result stating the (sufficient) conditions on the number of frequency samples required for sparse signal recovery from a subset of its DFT coefficients is given by the following theorem:

Theorem 3.1. *(Candès et al., 2006a; Donoho, 2006a) Let* $\mathbf{x} \in \mathbb{C}^N$ *be a sparse vector supported on a set* $K \subset \mathbb{Z}_N$, *and let* $S \subset \mathbb{Z}_N$ *be a subset of samples in the Fourier domain, chosen uniformly at random. Then, for a given constant* $\beta > 0$, *with probability* $p > 1 - O(N^{-\beta})$, *the solution to the optimization problem 3.2 is unique and equal to* $\mathbf{x}$ *if*

$$|S| \geq C_\beta |K| \log N, \tag{3.3}$$

where C_β *is approximately* $23(\beta + 1)$.

Essentially, the theorem states that, if a signal $\mathbf{x} \in \mathbb{C}^N$ is k-sparse, then it can be restored precisely from almost any random subset of its Fourier spectrum of size proportional to the sparsity k and to $\log N$. It is important to note that, since the above result is probabilistic, there exist particular signals and particular subsets of frequencies that do not allow for such recovery; some counter-examples are mentioned in (Candès et al., 2006a) and discussed later in this chapter.

The proof of the above theorem comprises the following two parts:

Deterministic part: If the restriction of the DFT matrix on a given subset of frequencies (i.e., a subset of rows) is close to being isometric (i.e., satisfies the so-called Restricted Isometry Property, or RIP) on subspaces of dimension $3|K|$, then the original signal can be recovered exactly.

Probabilistic part: The RIP property holds for *typical* subsets of the spectrum (i.e., typical subsets of rows of the DFT matrix). A "typical" subset denotes here a subset of fixed size k, selected uniformly at random from all possible size-k subsets of a given finite set, *with some (specified) high probability*.

We will now focus on the deterministic part and discuss sufficient conditions for RIP property, such as mutual coherence, *spark*, and null space properties, followed by proof of the exact signal recovery under RIP. Also, as we discuss below, the result of Theorem 3.1 can be extended to the general class of design matrices, beyond DFT, provided that some sufficient conditions, such as RIP, are satisfied.

3.4   Mutual Coherence

The term *mutual coherence* has several meanings, depending on the field of application. For example, in optics it describes the auto-correlation of a wave/field at different points with measurement time shift. Herein, we consider the linear-algebraic notion of the mutual coherence, shown to be one of the key properties in theoretical work on compressed sensing (Donoho and Huo, 2001; Tropp, 2006; Donoho et al.,

2006; Elad, 2010). Intuitively, mutual coherence characterizes the level of dependence among the columns of a matrix. Formally,

Definition 2. *(Mutual coherence) Given an $M \times N$ matrix $\mathbf{A}$, its* mutual coherence $\mu(\mathbf{A})$ *is defined as the maximal absolute inner product between pairs of its normalized columns, i.e.*

$$\mu(\mathbf{A}) = \max_{i,j \in \mathbb{Z}_N, \, i \neq j} \frac{|\mathbf{a}_i^* \mathbf{a}_j|}{||\mathbf{a}_i||_2 \, ||\mathbf{a}_j||_2}, \tag{3.4}$$

where $\mathbf{a}_i$ denotes the i-th column of the matrix $\mathbf{A}$, and $\mathbf{a}_i^*$ denotes its (conjugate) transpose (or simply transpose, in case of real-valued vectors).

Clearly, $\mu(\mathbf{A}) = 0$ if all columns of an $M \times N$ matrix $\mathbf{A}$ are mutually orthogonal. In the case when $N > M$, mutual coherence is strictly positive: $\mu(\mathbf{A}) > 0$.

In general, an achievable lower bound for mutual coherence is given by the *Welch bound* (Welch, 1974) (see also (Strohmer and Heath, 2003) for more recent applications and references). Note that equality in the Welch bound is obtained for a class of matrices called *tight frames*, or *Grassmanian frames*.

Theorem 3.2. *(Welch bound (Welch, 1974)) Let $\mathbf{A}$ be an $M \times N$ matrix, $M \leq N$, with normalized columns (i.e., columns having unit l_2-norm). Then*

$$\mu(\mathbf{A}) \geq \sqrt{\frac{N - M}{M(N - 1)}}. \tag{3.5}$$

Proof. Consider the Gram matrix $\mathbf{G} = \mathbf{A}^* \mathbf{A}$ of the matrix $\mathbf{A}$, with the elements $\mathbf{g}_{ij} = (\mathbf{a}_i^* \mathbf{a}_j)_{i,j \in \mathbb{Z}_N}$. The matrix $\mathbf{G}$ is positive semidefinite and self-adjoint. Since the rank of $\mathbf{A}$ does not exceed its smaller dimension, M, so does the rank of $\mathbf{G}$. Thus, $\mathbf{G}$ has no more than M nonzero eigenvalues (see section A.1.1 in Appendix for a short summary of eigentheory). Let $\{\lambda_l\}_{l \in \mathbb{Z}_r}, r \leq M$ be the set of nonzero eigenvalues of $\mathbf{G}$, counting multiplicity (i.e., if eigenvalue λ_l has k_l eigenvectors, it is included into the set k_l times). The trace of $\mathbf{G}$ is $Tr(\mathbf{G}) = \sum_{l \in \mathbb{Z}_r} \lambda_l$. Hence, using the Cauchy-Schwarz inequality, we get

$$Tr^2(G) = (\sum_{l \in \mathbb{Z}_r} \lambda_l)^2 \leq r \sum_{l \in \mathbb{Z}_r} \lambda_l^2 \leq M \sum_{l \in \mathbb{Z}_r} \lambda_l^2. \tag{3.6}$$

Since $Tr(\mathbf{G}^2) = Tr(\mathbf{G}\mathbf{G}) = \sum_{i,j \in \mathbb{Z}_N} |(\mathbf{a}_i^* \mathbf{a}_j)|^2 = \sum_{l \in \mathbb{Z}_r} \lambda_l^2$, we obtain

$$\frac{Tr^2(\mathbf{G})}{M} \leq \sum_{i,j \in \mathbb{Z}_N} |(\mathbf{a}_i^* \mathbf{a}_j)|^2.$$

Since the columns of $\mathbf{A}$ are normalized, the diagonal entries of $\mathbf{G}$, i.e. $\mathbf{g}_{ii} = (\mathbf{a}_i^* \mathbf{a}_i)_{i \in \mathbb{Z}_N}$, are all ones, which gives us $Tr(\mathbf{G}) = N$. Therefore, from the above inequality, we get:

$$\sum_{i,j \in \mathbb{Z}_N} |(\mathbf{a}_i^* \mathbf{a}_j)|^2 = N + \sum_{i,j \in \mathbb{Z}_N, i \neq j} |(\mathbf{a}_i^* \mathbf{a}_j)|^2 \geq \frac{N^2}{M},$$

which implies

$$\sum_{i,j\in\mathbb{Z}_N, i\neq j} |(\mathbf{a}_i^*\mathbf{a}_j)|^2 \geq \frac{N(N-M)}{M}. \tag{3.7}$$

We will now upper-bound the sum on the left, replacing each of its $N(N-1)$ terms with the maximum term over all $i,j \in \mathbb{Z}_N, i \neq j$, as follows:

$$N(N-1)\max_{i,j\in\mathbb{Z}_N, i\neq j} |(\mathbf{a}_i^*\mathbf{a}_j)|^2 \geq \sum_{i,j\in\mathbb{Z}_N, i\neq j} |(\mathbf{a}_i^*\mathbf{a}_j)|^2.$$

Combining this inequality with the one in eq. 3.7, we get

$$\max_{i,j\in\mathbb{Z}_N, i\neq j} |(\mathbf{a}_i^*\mathbf{a}_j)|^2 \geq \frac{N(N-M)}{M} \cdot \frac{1}{N(N-1)} = \frac{N-M}{M(N-1)}.$$

Since $\max_{i,j\in\mathbb{Z}_N, i\neq j} |(\mathbf{a}_i^*\mathbf{a}_j)|^2 = (\max_{i,j\in\mathbb{Z}_N, i\neq j} |(\mathbf{a}_i^*\mathbf{a}_j)|)^2 = \mu(\mathbf{A})^2$, we obtain the required inequality 3.5, which concludes the proof. $\square$

Note that the mutual coherence is easy to compute, as it requires only $O(NM)$ operations, unlike the two other matrix properties, RIP and *spark*, which will be introduced in the next sections. Both RIP and *spark* are NP-hard to compute (see, for example, section A.3 in Appendix and (Muthukrishnan, 2005) for the RIP computational complexity).

We will next introduce the notion of *spark*, and use it as a sufficient condition that guarantees the exact recovery of a sparse signal, i.e. the uniqueness of the solution to the problem in 3.2. Next we show a connection between the spark and the mutual coherence, and state the exact recovery result in terms of the latter notion.

3.5 Spark and Uniqueness of (P_0) Solution

Given a matrix $\mathbf{A}$, its *rank*, denoted $rank(\mathbf{A})$, is defined as the maximal number of linearly independent columns, and is a standard notion in linear algebra used for many years. On the other hand, the notions of *Kruskal's rank* and *spark* of a matrix $\mathbf{A}$, important for the analysis of sparse signal recovery, were introduced relatively recently.

Definition 3. *(Spark (Donoho and Elad, 2003)) Given an $M \times N$ matrix $\mathbf{A}$, its spark $spark(\mathbf{A})$, is defined as the minimal number of linearly dependent columns.*

Spark was used by (Gorodnitsky and Rao, 1997) to establish the uniqueness of sparse solution for 2.10, and was further developed by (Donoho and Elad, 2003). Note that spark is closely related to the Kruskal's rank (Kruskal, 1977), $krank(\mathbf{A})$, defined as *the maximal number k such that every subset of k columns of the matrix A is linearly independent.* It is easy to see that

$$spark(\mathbf{A}) = krank(\mathbf{A}) + 1. \tag{3.8}$$

Also, note that $rank(\mathbf{A}) \geq krank(\mathbf{A})$.

Although the notion of spark may look like a simple complement to the notion of rank, it actually requires evaluation of all possible subsets of columns of size up to $spark(\mathbf{A}) + 1$, which makes spark NP-hard to compute. However, in some cases it is easy to compute spark. For example, if the entries of $M \times N$ matrix $\mathbf{A}$, where $M \leq N$, are independent random variables with continuous density function(s), then, with probability one, any submatrix of size $M \times M$ has maximal rank $rank(\mathbf{A}) = M$, and hence its $spark(\mathbf{A}) = M + 1$.

We will now use spark to establish the following sparse signal recovery result:

Theorem 3.3. *((P_0) solution uniqueness via spark (Gorodnitsky and Rao, 1997; Donoho and Elad, 2003)) A vector $\bar{\mathbf{x}}$ is the unique solution of the problem (P_0) (stated in eq. 2.7),*

$$(P_0): \quad \min_{\mathbf{x}} ||\mathbf{x}||_0 \text{ subject to } \mathbf{y} = \mathbf{A}\mathbf{x}$$

if and only if $\bar{\mathbf{x}}$ is a solution of $\mathbf{A}\mathbf{x} = \mathbf{y}$ and $||\bar{\mathbf{x}}||_0 < spark(\mathbf{A})/2$.

Proof. Part 1: sufficient condition (Gorodnitsky and Rao, 1997). Let us assume that $\mathbf{x} \neq \bar{\mathbf{x}}$ is another solution of $\mathbf{A}\mathbf{x} = \mathbf{y}$. Then $\mathbf{A}(\bar{\mathbf{x}} - \mathbf{x}) = 0$, i.e., the columns of $\mathbf{A}$ corresponding to nonzero entries of the vector $\bar{\mathbf{x}} - \mathbf{x}$ are linearly dependent. Thus, the number of such columns, $||\bar{\mathbf{x}} - \mathbf{x}||_0$, must be greater or equal to $spark(\mathbf{A})$, by definition of spark. Since the support (the set of nonzeros) of $\bar{\mathbf{x}} - \mathbf{x}$ is a union of supports of $\bar{\mathbf{x}}$ and $\mathbf{x}$, we get $||\bar{\mathbf{x}} - \mathbf{x}||_0 \leq ||\bar{\mathbf{x}}||_0 + ||\mathbf{x}||_0$. But since $||\bar{\mathbf{x}}||_0 < spark(\mathbf{A})/2$, we get

$$||\mathbf{x}||_0 \geq ||\bar{\mathbf{x}} - \mathbf{x}||_0 - ||\bar{\mathbf{x}}||_0 > spark(\mathbf{A})/2,$$

which proves that $\bar{\mathbf{x}}$ is indeed the sparsest solution.

Part 2: necessary condition (Donoho and Elad, 2003). Note that, conversely, the uniqueness of a k-sparse solution of (P_0) implies that $k < spark(\mathbf{A})/2$. Indeed, assume that there exists a nonzero null vector $\mathbf{h}$ of $\mathbf{A}$ corresponding to $spark(\mathbf{A}) \leq 2k$ or with support not exceeding $2k$. Then there exist a couple of vectors $\bar{\mathbf{x}}, \mathbf{x}$ with support not exceeding k and $\mathbf{h} = \bar{\mathbf{x}} - \mathbf{x}$, and hence $\mathbf{A}\bar{\mathbf{x}} = \mathbf{A}\mathbf{x}$. If support of $\mathbf{h}$ is 1, then we can just take $\bar{\mathbf{x}} = 2\mathbf{x} = 2\mathbf{h}$; if support of $\mathbf{h}$ is greater that 1, then we can take $\bar{\mathbf{x}}, \mathbf{x}$ with non-intersecting support. This contradicts uniqueness of sparse solution of order $k \geq spark(\mathbf{A})/2$. $\qquad \square$

While spark is a useful notion for proving the above exact recovery result, it is hard to compute, as we discussed above. On the other hand, mutual coherence is easier to compute, and thus can serve as a more convenient tool in our analysis, once we establish the relationship between the two concepts. The next statement provides such connection, using mutual coherence to lower-bound the spark. (Recall that $\mu(\mathbf{A}) > 0$ when not all columns of $\mathbf{A}$ are mutually orthogonal, for example, in the case of $M < N$.)

Theorem 3.4. *(Spark and mutual coherence (Donoho and Elad, 2003)) For any* $M \times N$ *real-valued matrix* $\mathbf{A}$ *having* $\mu(\mathbf{A}) > 0$

$$spark(A) > 1 + \frac{1}{\mu(A)}. \tag{3.9}$$

Proof. First, we will normalize the columns of $\mathbf{A}$, obtaining the new matrix $\mathbf{A}'$, where $\mathbf{a}'_i = \mathbf{a}_i / \|\mathbf{a}_i\|_2$, for $1 \leq i \leq N$. Note that column normalization does not change the mutual coherence and spark properties of $\mathbf{A}$, and thus will not affect the statement of the theorem. Let λ be an eigenvalue of $\mathbf{A}$, and let $\mathbf{x}$ be the associated (nonzero) eigenvector, i.e. $\mathbf{Ax} = \lambda\mathbf{x}$. Then for the i-th coordinate $\mathbf{x}_i$, we get

$$(\lambda - \mathbf{a}_{ii})\mathbf{x}_i = \sum_{j \neq i} \mathbf{a}_{ij}\mathbf{x}_j. \tag{3.10}$$

Let i correspond to the largest coordinate of $\mathbf{x}$, i.e. $|\mathbf{x}_i| = \max_{j \in \mathbb{Z}_N} |\mathbf{x}_j|$. Note that $|\mathbf{x}_i|$ is strictly positive since $\mathbf{x}$ is nonzero. From eq. 3.10, we obtain the following inequality, known as Gershgorin's disks theorem:

$$|\mathbf{a}_{ii} - \lambda| \leq \sum_{j \neq i} |\mathbf{a}_{ij}| \left|\frac{\mathbf{x}_j}{\mathbf{x}_i}\right| \leq \sum_{j \neq i} |\mathbf{a}_{ij}|. \tag{3.11}$$

We now return to the spark estimate via the mutual coherence. By definition of spark, $k = spark(\mathbf{A})$ is the smallest number of dependent columns; let us take the set K of these columns and form a minor $\mathbf{A}|_K$ of the matrix $\mathbf{A}$ by restricting it to the columns in K. Clearly, $k = |K| = spark(\mathbf{A})$, and $spark(\mathbf{A}|_K) = k$. Let us consider the Gram matrix of $\mathbf{A}|_K$, i.e. $\mathbf{G} = (\mathbf{A}|_K)^*\mathbf{A}|_K$. The matrix $\mathbf{G}$ is degenerate, or singular, since $\mathbf{A}|_K$ is degenerate. Hence the spectrum (the set of all eigenvalues) of $\mathbf{G}$ contains zero, $0 \in Sp(\mathbf{G})$ (see A.5 in Appendix). By applying the inequality in 3.11 to $\mathbf{G}$ and eigenvalue $\lambda = 0$, we get

$$|1 - 0| \leq \sum_{i \neq j} |\mathbf{g}_{ij}| = \sum_{i \neq j} |\mathbf{a}_i^*\mathbf{a}_j| \leq (k-1)\mu(\mathbf{A}), \tag{3.12}$$

and

$$1 \leq (spark(\mathbf{A}) - 1)\mu(\mathbf{A}), \tag{3.13}$$

which implies the bound in 3.9.   □

Combining the above result with Theorem 3.3, we obtain the following sufficient condition for the exact signal recovery based on mutual coherence:

Theorem 3.5. *((P_0) solution uniqueness via mutual coherence (Donoho and Elad, 2003)) If $\bar{\mathbf{x}}$ is a solution of $\mathbf{Ax} = \mathbf{y}$ and $\|\bar{\mathbf{x}}\|_0 < 0.5(1 + \frac{1}{\mu(A)})$, then $\bar{\mathbf{x}}$ is the sparsest solution, i.e. it is the unique solution of the problem (P_0) in eq. 2.7.*

3.6 Null Space Property and Uniqueness of (P_1) Solution

So far we discussed the l_0-norm recovery, i.e. finding the sparsest solution. We gave the conditions for the uniqueness of the solution of (P_0) problem, based on the notion of spark as a necessary and sufficient condition, and the notion of mutual coherence as a sufficient condition. We will now consider the exact sparse recovery (i.e., the uniqueness of the solution) based on l_1-norm minimization as stated in the optimization problem (P_1) given by eq. 2.10, and focus on a condition that is both sufficient and necessary for such recovery.

Definition 4. *(Null space property (Cohen et al., 2009)) Given an $M \times N$ matrix $\mathbf{A}$, we say that $\mathbf{A}$ satisfies the* null-space property *of order k, $NSP(k)$, if, for any subset $K \subset \mathbb{Z}_N$ of size k and for any nonzero vector in the null-space of $\mathbf{A}$, $v \in Ker(\mathbf{A})$, the following inequality holds:*

$$||v|_K||_1 < ||v|_{K^c}||_1, \qquad (3.14)$$

where $v|_K$ and $v|_{K^c}$ are the restrictions of v on K and its complement K^c, respectively.

The following theorem due to Gribonval and Nielsen provides necessary and sufficient condition for the exact l_1 recovery.

Theorem 3.6. *((P_1) solution uniqueness via null space property (Gribonval and Nielsen, 2003)) A k-sparse solution $\mathbf{x}$ of the linear system $\mathbf{Ax} = \mathbf{y}$ is exactly recovered by solving the l_1-optimization problem (P_1) (stated in eq. 2.10):*

$$(P_1): \quad \min_{\mathbf{x}} ||\mathbf{x}||_1 \ subject\ to\ \mathbf{y} = \mathbf{Ax},$$

if and only if $\mathbf{A}$ satisfies NSP(k).

Proof. Part 1: NSP(k) implies uniqueness of (P_1) solution. Let us assume that $\mathbf{A}$ has a null space property of order k and $\bar{\mathbf{x}}$ is a solution of equation $\mathbf{Ax} = \mathbf{y}$ with support on the set K of size not exceeding k. Let $\mathbf{z}$ be another solution of this equation, i.e., $\mathbf{Az} = \mathbf{y}$, and thus $\mathbf{A\bar{x}} = \mathbf{Az}$. Hence, $\mathbf{v} = \mathbf{x} - \mathbf{z} \in Ker(\mathbf{A})$. Note that $\mathbf{x}_{K^c} = 0$, and therefore $\mathbf{v}|_{K^c} = \mathbf{z}|_{K^c}$. Then

$$||\bar{\mathbf{x}}||_1 \leq ||\bar{\mathbf{x}} - \mathbf{z}|_K||_1 + ||\mathbf{z}|_K||_1 = ||\mathbf{v}|_K||_1 + ||\mathbf{z}|_K||_1 <$$
$$||\mathbf{v}|_{K^c}||_1 + ||\mathbf{z}|_K||_1 = ||\mathbf{z}|_{K^c}||_1 + ||\mathbf{z}|_K||_1 = ||\mathbf{z}||_1. \qquad (3.15)$$

In other words, vector $\bar{\mathbf{x}}$ has strictly smaller l_1-norm than any other solution, and hence it is the only solution of the above optimization problem.

Part 2: Uniqueness of (P_1) solution implies NSP(k). Let us assume that, for a given $\mathbf{A}$ and given k, and for any given $\mathbf{y}$, a k-sparse solution of the (P_1) problem is always unique. We will now show that $\mathbf{A}$ has the NSP(k) property. Let $\mathbf{v}$ be a nonzero vector from kernel of $\mathbf{A}$, and let K be a subset of $\mathbb{Z}_N$ of size k. Since $\mathbf{v}|_K$ is k-sparse, and is a solution of the equation $\mathbf{A}\mathbf{z} = \mathbf{A}(\mathbf{v}|_K)$, then, by the above assumption, $\mathbf{v}|_K$ must be the unique l_1-norm minimizing solution of that equation. Since $\mathbf{A}(\mathbf{v}|_K + \mathbf{v}|_{K^c}) = 0$, we have $\mathbf{A}(-\mathbf{v}|_{K^c}) = \mathbf{A}\mathbf{v}|_K$. Note that $\mathbf{v}|_{K^c} + \mathbf{v}|_K \neq 0$, and thus $-\mathbf{v}|_{K^c} \neq \mathbf{v}|_K$. Then the vector $-\mathbf{v}|_{K^c}$ is another solution of the linear equation $\mathbf{A}\mathbf{z} = \mathbf{A}(\mathbf{v}|_K)$, and hence, due to the uniqueness assumption, its l_1-norm must be strictly higher than the l_1-norm of $\mathbf{v}|_K$, which is exactly the NSP(k) property.   □

Note that although the NSP is NP-hard to verify, it gives a nice geometric characterization of the exact recovery property.

3.7 Restricted Isometry Property (RIP)

As we already discussed, accurate recovery of a sparse signal depends on the properties of the set of measurements defined by the matrix $\mathbf{A}$. We will now consider a commonly used sufficient condition for the exact sparse recovery based on l_1-norm minimization called the *restricted isometry property*, or RIP. An attractive property of the RIP condition is that it provably holds for typical random matrices $\mathbf{A}$, such as matrices with i.i.d. random entries drawn from a wide variety of possible probability distributions. Essentially, RIP at the sparsity level k, or k-restricted isometry property, means that every subset of columns of $\mathbf{A}$ with cardinality less than k behaves very close to an isometric transformation, i.e. a transformation that preserves distances. Restricting near-isometry to subsets of k columns essentially means that the transformation will almost preserve the length of the corresponding sparse signals. Formally, following (Candès and Tao, 2005):

Definition 5. *(Restricted Isometry Property) The k-restricted isometry constant δ_k of the matrix $\mathbf{A}$ is the smallest quantity such that*

$$(1 - \delta_k)||\mathbf{x}||_2^2 \leq ||\mathbf{A}\mathbf{x}||_2^2 \leq (1 + \delta_k)||\mathbf{x}||_2^2 \qquad (3.16)$$

for all k-sparse vectors $\mathbf{x}$. The matrix $\mathbf{A}$ is said to satisfy the k-restricted isometry property, RIP(k), if there exists such constant δ_k that eq. 3.16 is satisfied.

The following lemma describes some simple properties of restricted isometry constant δ_k.

Lemma 3.7. *Let matrix $\mathbf{A}$ satisfy RIP(k). Then*

i) $\delta_1 \leq \delta_2 \leq \delta_3 \leq \dots$.

ii) *The restricted isometry constant δ_k may be evaluated as l_2-norm distortion on vectors with support of size k:*

$$\delta_k = \max_{K \subset \mathbb{Z}_N, |K| \leq k} ||\mathbf{A}|_K^* \mathbf{A}|_K - \mathbf{I}||_2 =$$

$$\sup_{||\mathbf{x}||_2 = 1, ||\mathbf{x}||_0 = k, K = supp(\mathbf{x})} |((\mathbf{A}|_K^* \mathbf{A}|_K - \mathbf{I})\mathbf{x})^* \mathbf{x}|,$$

where $\mathbf{I}$ is the identity matrix of size k.

Proof. Part i) follows from definition of RIP, since $(k-1)$-sparse vector is also k-sparse. Part ii) follows from the equalities

$$|\,||\mathbf{Ax}||_2^2 - ||\mathbf{x}||_2^2\,| = |((\mathbf{A}^*\mathbf{A} - \mathbf{I})\mathbf{x})^*\mathbf{x}| \quad \text{and} \quad ||\mathbf{B}||_2^2 = sup_{||\mathbf{x}||_2=1}|(\mathbf{B}^*\mathbf{Bx})^*\mathbf{x}|.$$

$\square$

The next lemma connects the restricted isometry constant with mutual coherence.

Lemma 3.8. *(RIP and mutual coherence) Let* $\mathbf{A}$ *be a matrix with* l_2*-norm normalized columns. Then*

i) $\mu(\mathbf{A}) = \delta_2$.

ii) *The restricted isometry constant* $\delta_k \leq (k-1)\mu(\mathbf{A})$.

Proof. i). By using the characteristic of the restricted isometry constant δ_2 from Lemma 3.7 ii), applied for $k = 2$ and $K = \{i, j\}$:

$$\mathbf{A}|_K^*\mathbf{A}|_K - \mathbf{I} = \begin{pmatrix} 0 & \mathbf{a}_i^*\mathbf{a}_j \\ \mathbf{a}_j^*\mathbf{a}_i & 0 \end{pmatrix} = \mathbf{a}_i^*\mathbf{a}_j \begin{pmatrix} 0 & 1 \\ 1 & 0 \end{pmatrix}, \tag{3.17}$$

where the last equality is due to the fact that $\mathbf{a}_j^*\mathbf{a}_i$ is real, the conjugate to $\mathbf{a}_j^*\mathbf{a}_i$ is $(\mathbf{a}_j^*\mathbf{a}_i)^* = \mathbf{a}_i^*\mathbf{a}_j$, and it is just its complex conjugate $(\mathbf{a}_j^*\mathbf{a}_i)^-$. Since $\mathbf{a}_j^*\mathbf{a}_i$ is real, then $\mathbf{a}_j^*\mathbf{a}_i = \mathbf{a}_i^*\mathbf{a}_j$. Hence $\delta_2 = \max_{i \neq j} ||\mathbf{A}|_K^*\mathbf{A}|_K - \mathbf{I}||_2 = \mu(\mathbf{A})$.

ii). Again, by using the presentation of δ_k from Lemma 3.7 ii) we have

$$\begin{aligned} \delta_k = \sup_{||\mathbf{x}||_2=1, ||\mathbf{x}||_0=k} |((\mathbf{A}|_K^*\mathbf{A}|_K - \mathbf{I})\mathbf{x})^*\mathbf{x}| \leq \\ \sup_{||\mathbf{x}||_2=1, ||\mathbf{x}||_0=k} (s-1)\mu(\mathbf{A})||\mathbf{x}||_2 = (s-1)\mu(\mathbf{A}). \end{aligned} \tag{3.18}$$

$\square$

3.8 Square Root Bottleneck for the Worst-Case Exact Recovery

In this section we give an example illustrating the fact that exact signal recovery (i.e., solution uniqueness) using mutual coherence in the worst-case scenario (i.e., for arbitrary matrix $\mathbf{A}$ with $N > M$ and nonzero, l_2-normalized columns) could not be done for a sparse signal with the support size larger than the order of $\sqrt{M}$. This scaling behavior is sometimes referred to as the *square root bottleneck*.

Indeed, if we consider Welch bound 3.6, then $\mu(A) > \sqrt{\frac{N-M}{M(N-1)}}$. For $M \leq N/2$ it gives $N - M \geq (N-1)/2$ and $\mu(A) > 1/\sqrt{2M}$ or $1/\mu(A) < \sqrt{2M}$. As it was mentioned above, this estimate is sharp on tight frames. Thus, for the worst-case scenario, the best estimate that we can obtain using spark estimate 3.9 will be an estimate not better than $spark(\mathbf{A}) > \sqrt{2M}$ without restriction on special form of matrices $\mathbf{A}$ with $N \geq 2M$, and such that it allows a k-sparse vector with $k < \sqrt{M/2}$ to be recovered exactly (see Theorem 3.3). This shows an order of magnitude for exact recovery in the worst-case scenario to be proportional to the square root of M.

Let us consider a special type of a sparse signal, sometimes called Dirac train (or Dirac comb), constructed as follows. Let $N = L^2$, and let a signal $\mathbf{v}$ equal 1 at coordinates that are multiples of L, up to L^2, and 0 at all other coordinates; in other words, $\mathbf{v}_i = \{1 \text{ for } i = lL, l = 1, ..., L\}$, and 0 otherwise. Note that the DFT of $\mathbf{v}$ coincides with $\mathbf{v}$. The restriction of the spectrum of $\mathbf{v}$ on the coordinates that are not multiples of L will be zero. Let us now consider the measurement matrix $\mathbf{A}$ that is the restriction of the DFT matrix on the subset of its $M = N - \sqrt{N}$ out of N rows corresponding to the above coordinates in the spectrum of $\mathbf{v}$. Clearly, the signal $\mathbf{v}$ cannot be recovered from these zero spectrum values, i.e., our L-sparse vector could not be recovered from equation $(\mathcal{F}(\mathbf{x}))|_K = 0$, where $K \subset \mathbb{Z}_N, K = \{n | n \neq lL\}$. In other words, we have $M = N - \sqrt{N}$ columns in $\mathbf{A}$, and $L = \sqrt{N} = O(\sqrt{M})$ nonzeros in a Dirac-comb vector that we cannot recover.

In summary, we showed that in order to guarantee sparse signal recovery in the case of an arbitrary $M \times N$ matrix $\mathbf{A}$, the signal must be "sufficiently" sparse, in a sense that its support size (number of nonzeros) must be on the order no larger than $\Omega(\sqrt{M})$. As we will discuss later, this square root bottleneck can be broken, i.e. a better scaling behavior can be obtained, if instead of a deterministic sparsity threshold we consider a probabilistic one, i.e., if we require the sparsity threshold to hold with high probability, rather than for all vectors $\mathbf{x}$. We will discuss next how RIP defined in the previous section implies the exact recovery, while leaving the probabilistic aspects for the next chapter.

3.9   Exact Recovery Based on RIP

This recent result in the series of estimates for noiseless sparse signal recovery is due to (Candès, 2008):

Theorem 3.9. *Let* $\mathbf{x}$ *and* $\mathbf{x}^*$ *denote solutions of the equation* $\mathbf{A}\mathbf{x} = \mathbf{y}$ *and the problem* $(P1)$*, respectively, and let* δ_k *of* $\mathbf{A}$ *denote the k-restricted isometry constant in definition 5 of Restricted Isometry Property. Let* $\mathbf{x}_k \in \mathbb{C}^N$ *denote the truncated version of* $\mathbf{x}$ *with all but the top k largest absolute values set to zero.*

I. *If* $\delta_{2k} < 1$*, and* $\mathbf{x}$ *is a k-sparse solution of* $\mathbf{A}\mathbf{x} = \mathbf{y}$*, then it is unique.*

II. *If* $\delta_{2k} < \sqrt{2} - 1$*, then*

      i) $||\mathbf{x}^* - \mathbf{x}||_1 \leq C_0 ||\mathbf{x} - \mathbf{x}_k||_1$

      ii) $||\mathbf{x}^* - \mathbf{x}||_2 \leq C_0 k^{-1/2} ||\mathbf{x} - \mathbf{x}_k||_1$,

where C_0 *is a constant.*

Theorem 3.9 II) implies that:

1. A k-sparse signal $\mathbf{x}$ can be recovered exactly from a collection of noiseless linear measurements $\mathbf{y} = \mathbf{Ax}$ by solving the l_1-norm minimization problem 2.10.

2. For arbitrary $\mathbf{x}$, the quality of its recovery depends on how well $\mathbf{x}$ is approximated by its k-sparse truncated version $\mathbf{x}_k$.

Proof of the Theorem 3.9. Part I. Let as assume that $\mathbf{Ax} = \mathbf{y}$ has two distinct k-sparse solutions, $\mathbf{x}_1$ and $\mathbf{x}_2$. Then $\tilde{\mathbf{x}} = \mathbf{x}_1 - \mathbf{x}_2$ is a $2k$-sparse vector. Using the condition $\delta_{2k} < 1$ and the lower bound in the definition of the RIP property, we get

$$0 < (1 - \delta_{2k})||\tilde{\mathbf{x}}||_2^2 \leq ||\mathbf{A}\tilde{\mathbf{x}}||_2^2, \tag{3.19}$$

which implies $\mathbf{A}\tilde{\mathbf{x}} = \mathbf{Ax}_1 - \mathbf{Ax}_2 > 0$ and thus contradicts our assumption $\mathbf{Ax}_1 = \mathbf{Ax}_2 = \mathbf{y}$.

Part II. We start with the lemma stating that a linear transformation satisfying the RIP property applied to vectors with disjoint support leaves them almost disjoint.

Lemma 3.10. *Let K, K' be a disjoint subset of $\mathbb{Z}_N$ and $|K| \leq k$, $|K'| \leq k'$. Let $\mathbf{A}$ be a transformation satisfying RIP properties with constant δ_k. Then for $\mathbf{x}, \mathbf{x}'$ with supports, K, K' accordingly holds:*

$$| < \mathbf{A}(\mathbf{x}), \mathbf{A}(\mathbf{x}') > | \leq \delta_{k+k'}||\mathbf{x}||_2||\mathbf{x}'||_2. \tag{3.20}$$

Proof. Without loss of generality we will assume that $\mathbf{x}$ and $\mathbf{x}'$ are unit vectors (otherwise, we can normalize them by their l_2-norms). Then RIP and disjointness of $\mathbf{x}$ and $\mathbf{x}'$ implies that

$$2(1 - \delta_{k+k'}) = (1 - \delta_{k+k'})||\mathbf{x} + \mathbf{x}'||_2^2 \leq ||\mathbf{A}(\mathbf{x} + \mathbf{x}')||_2^2$$
$$\leq (1 + \delta_{k+k'})||\mathbf{x} + \mathbf{x}'||_2^2 = 2(1 + \delta_{k+k'}). \tag{3.21}$$

Similar inequality is valid for $\mathbf{x} - \mathbf{x}'$:

$$2(1 - \delta_{k+k'}) \leq ||\Phi(\mathbf{x} - \mathbf{x}')||_2^2 \leq 2(1 + \delta_{k+k'}). \tag{3.22}$$

Using the parallelogram identity

$$\frac{1}{4}(||\mathbf{u} + \mathbf{v}||_2^2 + ||\mathbf{u} - \mathbf{v}||_2^2) = < \mathbf{u}, \mathbf{v} >, \tag{3.23}$$

combined with 3.21 and 3.22, we get

$$| < \mathbf{A}(\mathbf{x}), \mathbf{A}(\mathbf{x}') > | = \frac{1}{4}|||\mathbf{A}(\mathbf{x} + \mathbf{x}')||_2^2 - ||\mathbf{A}(\mathbf{x} - \mathbf{x}')||_2^2|$$
$$\leq \frac{1}{4}(2(1 + \delta_{k+k'}) - 2(1 - \delta_{k+k'})) = \delta_{k+k'}. \tag{3.24}$$

$\square$

We will first prove section ii) of Part II. Let us decompose $\mathbf{x}^*$ as $\mathbf{x}^* = \mathbf{x} + \mathbf{h}$. For the vector h we enumerate coordinates of $\mathbb{C}^N$ in decreasing order with respect to $\mathbf{h}$ so that $|\mathbf{h}_i| \geq |\mathbf{h}_j|$ for $i \leq j$. Denote by $T_i = [ik, (i+1)k - 1]$ intervals of indices of length k. (Note that the last interval may have a length less than k; however, this is not essential for the proof.) We will denote by $\mathbf{h}_{T_i}$ the restriction of $\mathbf{h}$ on T_i. Also, we will denote by $\mathbf{h}_{T_0 \cup T_1}$ the restriction of $\mathbf{h}$ on $T_0 \cup T_1$, and by $\mathbf{h}_{(I)^c}$ the restriction of $\mathbf{h}$ on indices complimentary to the indices in I.

Our goal now is to show that $||\mathbf{h}||_{l_1}$ is small, i.e., that $\mathbf{x}^*$, a solution of the l_1-norm optimization problem 2.10, is close to the solution $\mathbf{x}$ of $b = \mathbf{A}\mathbf{x}$.

We will use the following inequality:

$$||\mathbf{h}_{T_j}||_2 \leq \sqrt{\sum_{i=kj}^{(k+1)j-1} |\mathbf{h}_i|^2} \leq \sqrt{\sum_{i=kj}^{(k+1)j-1} ||\mathbf{h}_{T_j}||_\infty^2}$$
$$\leq k^{\frac{1}{2}} ||\mathbf{h}_{T_j}||_{l_\infty} \leq k^{\frac{1}{2}} k^{-1} \sum_{i=(k-1)j}^{kj-1} |\mathbf{h}_i|^2 = k^{-\frac{1}{2}} ||\mathbf{h}_{T_{j-1}}||_1. \tag{3.25}$$

The latter inequality is valid for $j \geq 1$. Summing 3.25 over all $j \geq 2$:

$$||\mathbf{h}_{(T_0 \cup T_1)^c}||_2 = ||\sum_{j \geq 2} \mathbf{h}_{T_j}||_2 \leq \sum_{j \geq 2} ||\mathbf{h}_{T_j}||_2$$
$$\leq \sum_{j \geq 1} k^{-\frac{1}{2}} ||\mathbf{h}_{T_j}||_1 = k^{-\frac{1}{2}} ||\mathbf{h}_{T_0^c}||_1. \tag{3.26}$$

Since $\mathbf{x} + \mathbf{h}$ is a minimal, then

$$||\mathbf{x}||_1 \geq ||\mathbf{x} + \mathbf{h}||_1 = \sum_{i \in T_0} |\mathbf{x}_i + \mathbf{h}_i| + \sum_{i \in T_0^c} |\mathbf{x}_i + \mathbf{h}_i|$$
$$\leq ||\mathbf{x}_{T_0}||_1 - ||\mathbf{h}_{T_0}||_1 + ||\mathbf{h}_{T_0^c}||_1 - ||\mathbf{x}_{T_0^c}||_1. \tag{3.27}$$

Since $\mathbf{x}_{T_0^c} = \mathbf{x} - \mathbf{x}_s$,

$$||\mathbf{h}_{T_0^c}||_1 \leq ||\mathbf{h}_{T_0}||_1 + 2||\mathbf{x}_{T_0^c}||_1. \tag{3.28}$$

Applying 3.26 and then 3.28 to bound $||\mathbf{h}_{(T_0 \cup T_1)^c}||_2$,

$$||\mathbf{h}_{(T_0 \cup T_1)^c}||_2 \leq k^{-\frac{1}{2}} ||\mathbf{h}_{T_0^c}||_1 \leq k^{-\frac{1}{2}} (||\mathbf{h}_{T_0}||_1 + 2||\mathbf{x}_{T_0^c}||_1)$$
$$\leq ||\mathbf{h}_{T_0}||_2 + 2k^{-\frac{1}{2}} ||\mathbf{x} - \mathbf{x}_k||_1, \tag{3.29}$$

where the last inequality is due to the Cauchy-Schwartz inequality

$$||\mathbf{h}_{T_0}||_1 = \sum_{i < s} \mathbf{h}_i \cdot sign(\mathbf{h}_i) \leq ||\{sign(\mathbf{h}_i)\}_{i<k}||_2 \cdot ||\mathbf{h}_{T_0}||_2 = k^{\frac{1}{2}} ||\mathbf{h}_{T_0}||_2. \tag{3.30}$$

We now define $e_0 \equiv k^{-\frac{1}{2}} ||\mathbf{x} - \mathbf{x}_k||_1$. The next step is to bound $||\mathbf{h}_{(T_0 \cup T_1)^c}||_2$. Since both $\mathbf{x}$ and $\mathbf{x}^*$ are solutions of the equation $b = \mathbf{A}u$, then $\mathbf{A}\mathbf{h} = 0$ and $\mathbf{A}\mathbf{h}_{(T_0 \cup T_1)} = -\sum_{j \geq 2} \mathbf{A}\mathbf{h}_{T_j}$.

Hence

$$||\mathbf{A}\mathbf{h}_{(T_0 \cup T_1)}||_2^2 = - < \mathbf{A}\mathbf{h}_{(T_0 \cup T_1)}, \sum_{j \geq 2} \mathbf{A}\mathbf{h}_{T_j} > . \tag{3.31}$$

From Lemma 3.10 it follows that

$$| < \mathbf{A}\mathbf{h}_{(T_k)}, \mathbf{A}\mathbf{h}_{T_j} > | \leq \delta_{2k} ||\mathbf{h}_{T_k}||_2 ||\mathbf{h}_{T_j}||_2, \tag{3.32}$$

for $k = 1, 2$ and $j \geq 2$.

Since T_0 and T_1 are disjoint, we get

$$2||\mathbf{h}_{T_0 \cup T_1}||_2^2 = 2(||\mathbf{h}_{T_0}||_2^2 + ||\mathbf{h}_{T_1}||_2^2) \geq (||\mathbf{h}_{T_0}||_2 + ||\mathbf{h}_{T_1}||_2)^2. \tag{3.33}$$

The RIP property applied to the $\mathbf{h}_{T_0 \cup T_1}$, 3.31 and equality $\mathbf{A}\mathbf{h}_{T_0 \cup T_1} = \mathbf{A}\mathbf{h}_{T_0} + \mathbf{A}\mathbf{h}_{T_1}$ imply that

$$(1 - \delta_{2k})||\mathbf{h}_{T_0 \cup T_1}||_2^2 \leq ||\mathbf{A}\mathbf{h}_{T_0 \cup T_1}||_2^2 \leq \delta_{2k}(||\mathbf{h}_{T_0}||_2 + ||\mathbf{h}_{T_1}||_2) \sum_{j \geq 2} ||\mathbf{h}_{T_j}||_2$$

$$\leq \sqrt{2}\delta_{2k} ||\mathbf{h}_{T_0 \cup T_1}||_2 \sum_{j \geq 2} ||\mathbf{h}_{T_j}||_2. \tag{3.34}$$

Let us use the notation $\rho \equiv \sqrt{2}\delta_{2k}(1 - \delta_{2k})^{-1}$. The assumption $\delta_{2k} < \sqrt{2} - 1$ in the formulation of Theorem 3.9 is equivalent to $\rho < 1$, which we use further.

Hence, 3.34 combined with 3.26 implies that

$$||\mathbf{h}_{T_0 \cup T_1}||_2 \leq \sqrt{2}\delta_{2k}(1 - \delta_{2k})^{-1} k^{-\frac{1}{2}} ||\mathbf{h}_{T_0} c||_1 \tag{3.35}$$

and, incorporating 3.29,

$$||\mathbf{h}_{T_0 \cup T_1}||_2 \leq \rho ||\mathbf{h}_{T_0 \cup T_1}||_2 + 2\rho e_0, \text{ or } ||\mathbf{h}_{T_0 \cup T_1}||_2 \leq 2\rho(1 - \rho)^{-1} e_0. \tag{3.36}$$

Summarizing,

$$||\mathbf{h}||_{l_2} \leq ||\mathbf{h}_{T_0 \cup T_1}||_2 + ||\mathbf{h}_{T_0 \cup T_1}^c||_2 \leq ||\mathbf{h}_{T_0 \cup T_1}||_2 + ||\mathbf{h}_{T_0 \cup T_1}||_2 + 2e_0$$

$$\leq 2(1 - \rho)^{-1}(1 + \rho)e_0. \tag{3.37}$$

That establishes part ii) of Theorem 3.9.

Part i) is based on the following consideration: The l_1 norm of $\mathbf{h}_{T_0}$ is estimated as

$$||\mathbf{h}_{T_0}||_1 \leq k^{\frac{1}{2}} ||\mathbf{h}_{T_0}||_2 \leq ||\mathbf{h}_{T_0 \cup T_1}||_2$$

$$\leq s^{\frac{1}{2}} \rho k^{-\frac{1}{2}} ||\mathbf{h}_{T_0}^c||_1 = \rho ||\mathbf{h}_{T_0}^c||_1. \tag{3.38}$$

Then since

$$||\mathbf{h}_{T_0}^c||_1 \leq \rho ||\mathbf{h}_{T_0}^c||_1 + 2||\mathbf{x}_{T_0}^c||_1, \tag{3.39}$$

then
$$||\mathbf{h}_{T_0}^c||_1 \le 2(1-\rho)^{-1}||\mathbf{x}_{T_0}^c||_1, \tag{3.40}$$

and hence

$$\begin{aligned}||\mathbf{h}||_1 = ||\mathbf{h}_{T_0}||_1 + ||\mathbf{h}_{T_0}^c||_1 &\le (\rho + 2(1-\rho)^{-1})||\mathbf{x}_{T_0}^c||_1 \\ &\le 2(1+\rho)(1-\rho)^{-1}||\mathbf{x}_{T_0}^c||_1.\end{aligned} \tag{3.41}$$

This establishes part i) of Theorem 3.9.   □

3.10   Summary and Bibliographical Notes

Once again, we would like to note that existing literature on the subject of sparse recovery is immense, and only a relatively small fraction of it was presented here. This chapter focused primarily on the seminal work by (Donoho, 2006a) and (Candès et al., 2006a) which gave rise to the compressed sensing field. In (Candès et al., 2006a), a puzzling empirical phenomenon was presented, which at a first glance appeared to contradict the conventional sampling theorem (Whittaker, 1990; Nyquist, 1928; Kotelnikov, 2006; Shannon, 1949), achieving accurate signal recovery with a much smaller number of samples than the sampling theorem would generally require. The "catch" here was that the sparse structure of a signal was exploited. Moreover, (Candès et al., 2006a) provided a considerable improvement over the state-of-the-art results on sparse signal recovery presented earlier in (Donoho and Stark, 1989) and (Donoho and Huo, 2001), by reducing the sample size required for sparse recovery from the square root to the logarithmic in the signal dimension.

Next, we considered some key properties of the design matrix that are essential in sparse recovery, such as mutual coherence, spark, and null space properties. The mutual coherence was considered in (Donoho and Huo, 2001; Tropp, 2006; Donoho et al., 2006; Elad, 2010). Spark was used to show uniqueness of the sparse solution in (Gorodnitsky and Rao, 1997). The Kruskal rank (or krank), a notion related to spark, was considered in (Kruskal, 1977). The relation between spark and mutual coherence was presented in (Donoho and Elad, 2003).

The null space property is a necessary and sufficient condition for the sparse recovery, and was presented in (Gribonval and Nielsen, 2003) and (Cohen et al., 2009). The square root bottleneck results are due to (Donoho and Stark, 1989; Donoho and Huo, 2001). The RIP property was introduced in (Candès et al., 2006a). Our exposition of the sparse signal recovery based on RIP follows the one presented in (Candès, 2008).

Chapter 4

Theoretical Results (Probabilistic Part)

In this chapter we give examples of matrices satisfying RIP. Namely, we consider matrices with i.i.d. random entries that decay sufficiently fast (e.g., subgaussian), matrices with randomly selected rows from the DFT (or DCT) matrix, and matrices with the rows randomly selected from a general orthogonal matrix. We provide complete and self-contained proofs of RIP in these cases, trying to keep them concise but complete. All necessary background material that goes beyond the standard probability and linear algebra courses can be found in the Appendix. Also, recall that matrices that have RIP satisfy this property uniformly for all subsets of size s or less. For non-uniform results that nevertheless imply the sparse signal recovery we refer to (Candès and Plan, 2011) and to recent monographs (Foucart and Rauhut, 2013; Chafai et al., 2012).

We would like to note that the material presented in this chapter is more advanced and may require a deeper mathematical background than the rest of this book. On the other hand, the proofs in the chapter are sufficiently independent from the other topics we cover, and thus skipping them in the first reading will not have a negative effect on the understanding of the remaining material.

4.1 When Does RIP Hold?

In this section, we introduce three examples of matrices satisfying RIP, which actually fall into the following two categories: the first example includes matrices with i.i.d. random entries following subgaussian distributions, while the other two examples involve large orthogonal (or DFT) matrices. In these examples we chose the rows randomly.

Example 1.   *1. Random matrices with i.i.d. entries (Candès and Tao, 2006; Donoho, 2006d; Rudelson and Vershynin, 2006). Let matrix* $\mathbf{A}$*'s entries be i.i.d. for a subgaussian distribution with* $\mu = 0$ *and* $\sigma = 1$*. Then* $\hat{\mathbf{A}} = \frac{1}{\sqrt{M}}\mathbf{A}$ *satisfies RIP with* $\delta_S \leq \delta$ *when* $M \geq const(\varepsilon, \delta) \cdot S \cdot \log(2N/S)$ *with probability* $p > 1 - \varepsilon$*. Distribution examples: Gaussian, Bernoulli, subgaussian.*

    2. Fourier ensemble (Candès and Tao, 2006; Rudelson and Vershynin, 2006). Let $\hat{\mathbf{A}} = \frac{1}{\sqrt{M}}\mathbf{A}$ *with* $\mathbf{A}$ *being* M *randomly selected rows from an* $N \times N$ *DFT matrix. Then* $\hat{\mathbf{A}}$ *satisfies RIP with* $\delta_S \leq \delta$ *providing* $M \geq const(\varepsilon, \delta) \cdot S \cdot \log^4(2N)$*, with probability* $p > 1 - \varepsilon$*.*

    3. General orthogonal ensembles (Candès and Tao, 2006). Let $\hat{\mathbf{A}}$ *be* M *randomly selected rows from an* $N \times N$ *orthonormal matrix* U *with re-normalized columns. Then* S*-sparse recover* $\mathbf{x}$ *with high probability when* $M \geq const \cdot \mathcal{M}^2(U) \cdot S \cdot \log^6 N$*.*

The rest of the chapter is devoted to proving these results.

4.2 Johnson-Lindenstrauss Lemma and RIP for Subgaussian Random Matrices

In this section we establish part 1 of example 1. While the following result does not give the best possible estimate, it has, however, a very clear and short proof. In our exposition we follow (Baraniuk et al., 2008).

The result we are about to present is referred to as the Johnson-Lindenstrauss lemma. Essentially, it states that $|Q|$ points in the Euclidean space R^N can be mapped into $\mathbb{R}^n$ so that the distances between the points are distorted by less than a multiplicative factor of $1 \pm \varepsilon$ and n is the order of $\ln(|Q|)/\epsilon^2$. Note that, in this chapter only, we will use a somewhat different notation for the norm of a vector, that explicitly includes the dimensionality of the space; namely, $\|\mathbf{x}\|_{\ell_2^N}$ will denote the l_2-norm of a vector $\mathbf{x} \in \mathbb{R}^N$.

Lemma 4.1. *(Johnson and Lindenstrauss, 1984) Let* $\epsilon \in (0, 1)$ *be given. For every set* Q *of* $|Q|$ *points in* R^N*, if* n *is an integer* $n > n_0 = O(\ln(|Q|))/\epsilon^2)$*, then there* *exists a Lipschitz mapping* $f : \mathbb{R}^N \to \mathbb{R}^n$ *such that*

$$(1 - \epsilon)\|\mathbf{u} - \mathbf{v}\|_{\ell_2^N}^2 \leq \|f(\mathbf{u}) - f(\mathbf{v})\|_{\ell_2^n}^2 \leq (1 + \epsilon)\|\mathbf{u} - \mathbf{v}\|_{\ell_2^N}^2 \qquad (4.1)$$

for all $\mathbf{u}, \mathbf{v} \in Q$.

The Johnson-Lindenstrauss lemma is a concentration inequality since its proof is based on the following statement.

Let $\Phi(\omega)$ be a random matrix with entries $\Phi_{ij} = \frac{1}{\sqrt{n}} R_{i,j}$, where $R_{i,j}$ are i.i.d. random variables with $\mathrm{E}[R_{ij}] = 0, \mathrm{Var}[R_{ij}] = 1$, and a uniform subgaussian tail defined by constant a. Once again, we refer the reader to the Appendix for definitions and other information from the theory of subgaussian random variables.

Theorem 4.2. *For any* $x \in \mathbb{R}^N$, *the random variable* $\|\Phi(\omega)\mathbf{x}\|_{\ell_2^n}^2$ *is strongly concentrated around its expected value,*

$$\mathrm{Prob}(|\|\Phi(\omega)\mathbf{x}\|_{\ell_2^n}^2 - \|\mathbf{x}\|_{\ell_2^N}^2| \geq \epsilon\|\mathbf{x}\|_{\ell_2^N}^2) \leq 2e^{-nc_0(\epsilon)}, \ 0 < \epsilon < 1, \qquad (4.2)$$

where the probability is taken over all $n \times N$ *matrices* $\Phi(\omega)$, *and* $c_0(\epsilon) > 0$ *is a constant depending only on* $\epsilon \in (0, 1)$.

The inequality given above in eq. 4.2 is called the *Johnson-Lindenstrauss concentration inequality*. Concentration inequalities form a subarea of discrete geometry. A reader interested in deeper understanding of this subject is referred to an introductory book by (Matoušek, 2002).

We are now ready to establish RIP for matrices with subgaussian random entries.

Theorem 4.3. *(RIP for random matrices) Suppose that* n, N, *and* $0 < \delta < 1$ *are given. If the probability distribution generating the* $n \times N$ *matrices* $\Phi(\omega)$, $\omega \in \Omega^{nN}$, *satisfies the Johnson-Lindenstrauss concentration inequality, then there exist constants* $c_1, c_2 > 0$ *depending only on* δ *such that RIP (3.16) holds for* $\Phi(\omega)$ *with the prescribed* δ *and any* $k \leq c_1 n / \log(N/k)$ *with probability* $\geq 1 - 2e^{-c_2 n}$.

The proofs of the theorems are given in the next sections.

4.2.1 Proof of the Johnson-Lindenstrauss Concentration Inequality

We now present the proof of the Johnson-Lindenstrauss concentration inequality and lemma, following (Matoušek, 2002).

Proof. By dividing the expression under the probability sign in 4.2 by $\|\mathbf{x}\|_{\ell_2^N}$, we may assume that $\|\mathbf{x}\|_{\ell_2^N} = 1$. Then

$$\|\Phi(\omega)\mathbf{x}\|_{\ell_2^n}^2 - 1 = \frac{1}{\sqrt{n}} \frac{1}{\sqrt{n}} \left(\sum_{i=1}^{n} \left(\sum_{j=1}^{N} R_{ij}\mathbf{x}_j \right)^2 - n \right), \qquad (4.3)$$

or $\|\Phi(\omega)\mathbf{x}\|_{\ell_2^n}^2 - 1$ is distributed as $\frac{1}{\sqrt{n}} Z$, where $Z = \frac{1}{\sqrt{n}} (\sum_{i=1}^{n} (Y_i)^2 - n)$, and

$Y_i = \sum_{j=1}^N R_{ij}\mathbf{x}_j$. Since R_{ij} are i.i.d., Y_i are independent. By Theorem A.6 (see Appendix), Y_i are subgaussian random variables with $E[Y_i] = 0, \mathrm{Var}[Y_i] = 1$.
Since $\|\mathbf{x}\|_{\ell_2^N} = 1$,

$$\mathrm{Prob}[\|\Phi(\omega)\| \geq 1 + \epsilon] \leq \mathrm{Prob}[\|\Phi(\omega)\|^2 \geq 1 + 2\epsilon] = \mathrm{Prob}[Z \geq 2\epsilon\sqrt{n}]. \quad (4.4)$$

Since we may choose $\epsilon \leq \dfrac{1}{2}$, the last probability by Proposition A.8 and Proposition A.3 does not exceed

$$e^{-a(2\epsilon\sqrt{n})^2} = e^{-4a\epsilon^2 n} \leq e^{-C(\epsilon)n}, \ C(\epsilon) = 4a\epsilon^2. \quad (4.5)$$

The proof of the estimate $\mathrm{Prob}[\|\Phi(\omega)\| \leq 1 - \epsilon] \leq e^{-C(\epsilon)n}$ is similar. $\quad\square$

The Johnson-Lindenstrauss lemma is a direct corollary of estimates 4.4 and 4.5

Proof (Johnson-Lindenstrauss lemma). Consider $|Q|^2$ vectors $\mathbf{u} - \mathbf{v}$, where $\mathbf{u}, \mathbf{v} \in Q$, and take arbitrary $F(\omega)$ with entries being subgaussian i.i.d. random variables with $E[R_{ij}] = 0, \mathrm{Var}[R_{ij}] = 1$, and a uniform subgaussian tail defined by constant a. By choosing $n > C\log(|Q|)/(a\epsilon^2)$, we get that

$$|Q|^2(\mathrm{Prob}[\|\Phi(\omega)\|^2 \geq 1 + \epsilon] + \mathrm{Prob}[\|\Phi(\omega)\|^2 \geq 1 - \epsilon]) < 2|Q|^2 e^{-4a\epsilon^2 n} < 1.$$

In other words, there exists ω_0, such that for every pair $\mathbf{u}, \mathbf{v} \in Q$,

$$(1 - \epsilon)\|\mathbf{u} - \mathbf{v}\|^2 \leq \|\Phi(\omega_0)(\mathbf{u} - \mathbf{v})\|^2 \leq (1 + \epsilon)\|\mathbf{u} - \mathbf{v}\|^2.$$

We choose $f = \Phi(\omega_0)$. $\quad\square$

4.2.2 RIP for Matrices with Subgaussian Random Entries

In order to proceed from the Johnson-Lindenstrauss lemma to RIP, we need to establish that the concentration inequality 4.2 holds uniformly for all unit vectors. Following the spirit of the Johnson-Lindenstraus lemma's proof, we need to show that, with large probability, (1) $\Phi(\omega)$ is bounded and (2) the estimate 4.2 holds on centers of small balls covering the unit sphere. In order to do that, we need to estimate the number of small balls covering the unit sphere.

We will first introduce the definition of the ϵ-cover of a body $D \subset \mathbb{R}^n$.

Definition 6. *Let $D \subset \mathbb{R}^N$ and let $\epsilon > 0$. The set $\mathcal{N} \subset D$ is called ϵ-net of the set D if the distance from every point of D to the set $\mathcal{N}$ is not greater than ϵ:*

$$\forall \mathbf{x} \in D \, \exists \mathbf{y} \in N \text{ with } dist(\mathbf{x}, \mathbf{y}) \leq \epsilon.$$

The minimal size of set $\mathcal{N}$ is called the covering number. The covering number for a pair K, D of convex bodies in $\mathbb{R}^n$, denoted $N(K, D)$, is defined as a minimal size of covering K with shifts of D.

Next we will estimate the covering number for unit sphere S^{n-1}.

Lemma 4.4. *(Size of ϵ-net of the sphere S^{n-1}) Let $0 < \epsilon < 1$. Then ϵ-net $\mathcal{N}$ may be chosen with*

$$|\mathcal{N}| \leq (1 + \frac{2}{\epsilon})^n. \tag{4.6}$$

Proof. The proof goes back to at least (Milman and Schechtman, 1986); see also (Rudelson and Vershynin, 2006). Instead of considering ϵ-net, we consider the set $\mathcal{N}'$ with the smallest distances between points at least ϵ, i.e. the so-called ϵ-*separated set*,

$$\mathcal{N}' = \{\mathbf{x}_i | dist(\mathbf{x}_i, \mathbf{x}_j) \geq \epsilon\}, \tag{4.7}$$

and consider the maximal such net. This net may be constructed just by adding points at the distance of ϵ to already chosen points. The set $\mathcal{N}'$ is an ϵ-net. Otherwise, there will be a point on S^{n-1} with the distance from $\mathcal{N}' > \epsilon$, and we can choose more points on the distance ϵ from $\mathcal{N}'$. Now we apply volume estimate arguments. Consider open unit balls $B(\mathbf{x}_i, \epsilon/2)$ with the origins at the points within the set $\mathcal{N}'$. Then the balls do not intersect, and they all are contained in the ball $B(0, 1 + \epsilon/2)$ of radius $1 + \frac{\epsilon}{2}$, centered at the origin. Hence, estimating volumes of the balls,

$$|\mathcal{N}'|(\frac{\epsilon}{2})^n \leq (1+\epsilon)^n, \tag{4.8}$$

establishes 4.6. □

With a small variation, the proof of Lemma 4.4 estimates $N(K, D)$.

Lemma 4.5. *(Volumetric estimate, (Milman and Schechtman, 1986; Rudelson, 2007)) Let $0 < \epsilon < 1$, and K, D are convex bodies in $\mathbb{R}^n$. Then D-net $\mathcal{N}$ on K may be chosen with*

$$|\mathcal{N}| \leq Volume(K + D)/Volume(D). \tag{4.9}$$

Proof. Let $\mathcal{N} = \{\mathbf{x}_1, ..., \mathbf{x}_N\} \subset K$ be a set with $\mathbf{x}_i + D \cap \mathbf{x}_j + D = \emptyset$ for $i \neq j; i, j \in \mathbb{Z}_N$. Then

$$Volume(K + D) \geq Volume(\bigcup_1^N (\mathbf{x}_i + D)) \tag{4.10}$$

$$= \sum_1^N Volume(\mathbf{x}_i + D) = N \cdot Volume(D). \tag{4.11}$$

Hence,

$$N(K, D) \leq N \leq \frac{Volume(K + D)}{Volume(D)}. \tag{4.12}$$

□

Corollary 4.6. *Let $K \subset \mathbb{R}^n$ be a convex body. Then for positive $\epsilon < 1$,*

$$N(K, \epsilon K) \leq (1 + \frac{1}{\epsilon})^n. \tag{4.13}$$

We now apply estimate 4.13 to the proof of uniform form of 4.2.

Theorem 4.7. *Let* $\Phi(\omega)$, $\omega \in \Omega^{nN}$, *be a random matrix of size* $n \times N$ *drawn according to any distribution that satisfies the concentration inequality 4.2. Then, for any set T with $|T| = k < n$ and any $0 < \delta < 1$, we have*

$$(1 - \delta)\|\mathbf{x}\|_{\ell_2^N} \leq \|\Phi(\omega)\mathbf{x}\|_{\ell_2^n} \leq (1 + \delta)\|\mathbf{x}\|_{\ell_2^N} \text{ for all } \mathbf{x} \in X_T \qquad (4.14)$$

with a probability at least

$$1 - 2(9/\delta)^k e^{-c_0(\delta/2)n}. \qquad (4.15)$$

Proof. Note that it is enough to establish 4.14 for x with $\|\mathbf{x}\|_2 = 1$. Next, we choose $\delta/4$-net Q in S^{k-1}. In other words, for any $\mathbf{x} \in S^{k-1}$ holds

$$dist(x, Q) \leq \delta/4. \qquad (4.16)$$

According to Lemma 4.4, the set Q may be chosen with a size not exceeding $(1 + 8/\delta)^n \leq (9/\delta)^n$. By applying estimate 4.2 to the union of points Q, we get with probability at least as in estimate 4.15 inequality

$$(1 - \delta/2)\|\mathbf{q}\|_{\ell_2^N}^2 \leq \|\Phi(\omega)\mathbf{q}\|_{\ell_2^n}^2 \leq (1 + \delta/2)\|\mathbf{q}\|_{\ell_2^N}^2 \text{ for all } \mathbf{q} \in Q, \qquad (4.17)$$

or by taking the square root,

$$(1 - \delta/2)\|\mathbf{q}\|_{\ell_2^N} \leq \|\Phi(\omega)\mathbf{q}\|_{\ell_2^n} \leq (1 + \delta/2)\|\mathbf{q}\|_{\ell_2^N} \text{ for all } \mathbf{q} \in Q. \qquad (4.18)$$

Let $B = \sup_{\mathbf{x} \in S^{k-1}} \|\Phi(\omega)x\| - 1$. Then $B \leq \delta$. Indeed, fix $\mathbf{x} \in S^{k-1}$. Pick a $\mathbf{q} \in Q$ satisfying $dist(\mathbf{x}, \mathbf{q}) \leq \delta/4$. Then

$$\|\Phi\mathbf{x}\|_{\ell_2^n} \leq \|\Phi(\omega)\mathbf{q}\|_{\ell_2^n} + \|\Phi(\omega)(\mathbf{x} - \mathbf{q})\|_{\ell_2^n} \leq 1 + \delta/2 + (1 + B)\delta/4. \qquad (4.19)$$

By taking the supremum over all $\mathbf{x} \in S^{n-1}$ we get

$$B \leq \delta/2 + (1 + B)\delta/4, \qquad (4.20)$$

or $B \leq 3\delta/4/(1 - \delta/4) \leq \delta$. This establishes the estimate from above for 4.14. The estimate from below follows from

$$\|\Phi(\omega)\mathbf{x}\|_{\ell_2^n} \geq \|\Phi(\omega)\mathbf{q}\|_{\ell_2^n} - \|\Phi(\omega)(\mathbf{x} - \mathbf{q})\|_{\ell_2^n} \geq 1 - \delta/2 - (1 + \delta)\delta/4 \geq 1 - \delta. \qquad (4.21)$$

$\square$

Now we establish Theorem 4.3.

Proof. For each k-dimensional subspace X_k estimate 4.14 fails with probability at most

$$2(9/\delta)^k e^{-c_0(\delta/2)n}. \qquad (4.22)$$

For the fixed basis there are $\binom{N}{k} \leq (eN/k)^k$ such subspaces. Hence, for an arbitrary subspace, estimate 4.14 fails with probability at most

$$2(eN/k)^k(12/\delta)^k e^{-c_0(\delta/2)n} = 2e^{-c_0(\delta/2)n + k[\log(eN/k) + \log(12/\delta)]}. \qquad (4.23)$$

For a fixed $c_1 > 0$, whenever $k \leq c_1 n / \log(N/k)$, we will have that the exponent in the exponential on the right side of 4.23 is $\leq -c_2 n$, provided that $c_2 \leq c_0(\delta/2) - c_1[1 + (1 + \log(12/\delta))/\log(N/k)]$. Hence, we can always choose $c_1 > 0$ sufficiently small to ensure that $c_2 > 0$. Thus, with probability $1 - 2e^{-c_2 n}$ the matrix $\Phi(\omega)$ satisfies 4.14 for x with $\|supp(\mathbf{x})\|_{l_0} \leq k$. $\square$

4.3 Random Matrices Satisfying RIP

In this section we review several cases of random matrices satisfying RIP. We describe a few types of randomness, different decay requirements on the distributions (Gaussian, Bernoulli, subgaussian, subexponential, heavy tail), and i.i.d conditions on entries, rows, or columns.

Example 1 shows different types of "randomness" that can lead to RIP of a matrix. In one case, we have random entries of the matrix – see example 1(1). Note that the order of the estimate $S \log(N/S)$ in 1 cannot be improved, as it was shown in the recent work on RIP and phase transitions (see, for example, (Donoho and Tanner, 2009), as well as (Foucart et al., 2010; Garnaev and Gluskin, 1984)), since it is related to the lower bound of the Gelfand's width of l_1^N ball. For this type of randomness, it is possible to consider distributions with different decay, such as Gaussian, subgaussian, or subexponential. For a similar result with subgaussian entries, decay restriction relaxed to the subexponential entries decay, with estimate of order $S \log^2(N/S)$, see (Adamczak et al., 2011).

Other series of results, starting from (Candès et al., 2006a), relate to the randomly chosen rows of some matrix; see examples 1(2) and 1(3). Since we do not put any restrictions on the decay of the distribution, it is the case of the so-called *heavy-tailed* random matrices. One of the strongest results in this direction is due to (Rudelson and Vershynin, 2008), where for the constant probability maximal size of recoverable dimension of the vector is order of $S \log^4(N/S)$; see also (Rauhut, 2008; Foucart and Rauhut, 2013) for the result on the Fourier matrices. Recently, this estimate was improved to $S \log^3(N/S)$; see (Cheraghchi et al., 2013).

In the rest of the chapter we prove the result presented in (Rudelson and Vershynin, 2008), following the original version of the proof. We use an unpublished series of lectures by (Rudelson, 2007) for the self-contained proof of the Dudley inequality, needed to establish Uniform Rudelson inequality or Uniform Law of Large Numbers (ULLN); also, see Theorem 11.17 in (Ledoux and Talagrand, 2011) for a similar exposition of the Dudley inequality.

We proceed as follows. After formulating the main result, we illustrate it on a couple of examples. Next, we will formulate the uniform Rudelson inequality (URI). We will also need the statements on uniform deviation and uniform symmetrization. Finally, we prove the main result and conclude the chapter with the proof of the URI, including exposition of the Dudley inequality.

4.3.1 Eigenvalues and RIP

We can formulate RIP in terms of singular values of the matrix $\mathbf{A}$. Recall that *singular value* of $N \times n$ matrix $\mathbf{A}$ is a non negative real λ such that there exists a pair of vectors $\mathbf{v}, \mathbf{u}$ with $\mathbf{Av} = \lambda\mathbf{u}$ and $\mathbf{A'u} = \lambda\mathbf{v}$, where $\mathbf{A'}$ denotes the transpose of $\mathbf{A}$. An *SVD decomposition* is representation of $\mathbf{A} = \mathbf{U\Sigma V}$ as a product of orthogonal matrices $\mathbf{U}$ and $\mathbf{V}$ (i.e., $\mathbf{UU'} = \mathbf{I}_N$ and $\mathbf{U'U} = \mathbf{VV'} = \mathbf{V'V} = \mathbf{I}_n$), and a non-negative diagonal $n \times n$ matrix Σ. The diagonal elements of the matrix Σ are called singular values. Usually, singular values are listed in a decreasing order:

$$s_1 \geq s_2 \geq \dots \geq s_n \geq 0. \qquad (4.24)$$

Let $s_{min} = s_n$ and $s_{max} = s_1$ be the minimal and the maximal singular values, respectively. Note that $0 \leq s_{min} \leq s_{max} = ||\mathbf{A}||$. Due to SVD decomposition, $< \mathbf{Ax}, \mathbf{Ax} > = < \mathbf{A'Ax}, \mathbf{x} > = < \Sigma^2\mathbf{Vx}, \mathbf{Vx} >$, and hence $s_{min}^2 = \min_{|bx||_2=1} < \mathbf{Ax}, \mathbf{Ax} >$, and $s_{max}^2 = \max_{||\mathbf{x}||_2=1} < \mathbf{Ax}, \mathbf{Ax} >$. Thus, the RIP 3.16 condition may be rewritten in the form

$$(1 - \delta_k) \leq s_{min}^2 \leq s_{max}^2 \leq 1 + \delta_k. \qquad (4.25)$$

Recall that another way of presenting RIP 3.16 is as follows:

$$\delta_k = \max_{T \subset \mathbb{Z}_N, |T| \leq k} ||\mathbf{A}|'_T \mathbf{A}|_T - \mathbf{I}_{\mathbb{R}^{|T|}}||; \qquad (4.26)$$

see Lemma 3.7, ii).

Thus, a uniform estimate of the singular values over all k-dimensional restrictions of the matrix $\mathbf{A}$ on at most k coordinates T is given by

$$\max_{T \subset \{1,\dots,n\}; |T|} \{|1 - s_{max}(\mathbf{A}|_T)|, |1 - s_{min}(\mathbf{A}|_T)|\} \leq \delta, \qquad (4.27)$$

which implies that

$$1 - \delta^2 \leq s_{min}(\mathbf{A}|_T) = \min_{||\mathbf{x}||_2=1, supp(\mathbf{x}) \subset T} ||\mathbf{A}|_T \mathbf{x}||_2 \qquad (4.28)$$

$$\leq \max_{||\mathbf{x}||_2=1, supp(\mathbf{x}) \subset T} ||\mathbf{A}|_T \mathbf{x}||_2 = s_{max}(\mathbf{A}|_T) \leq 1 + \delta^2, \qquad (4.29)$$

or, in other words, RIP holds with $\delta_k \leq \delta^2$.

Given two vectors $\mathbf{x}$ and $\mathbf{y}$, we consider an $\mathbf{xy'}$ matrix with coefficients $(x_i y_j)|_{i,j}$. Sometimes we use tensor product notion $\mathbf{x} \otimes \mathbf{y}$ for the matrix $\mathbf{xy'}$.

4.3.2 Random Vectors, Isotropic Random Vectors

A random n-dimensional vector is given by some probability measure in $\mathbb{R}^n$. Expectation $E\mathbf{x}$ of a random vector $\mathbf{x}$ is the coordinate-wise expectation. The second moment of a random vector $\mathbf{x}$ is the matrix $\Sigma\mathbf{x} = E\mathbf{x}\mathbf{x}' = E\mathbf{x} \otimes \mathbf{x}$. Recall that the *covariance of random vector* $\mathbf{x}$ is

$$cov(\mathbf{x}) = E(\mathbf{x} - E\mathbf{x})(\mathbf{x} - E\mathbf{x})' \tag{4.30}$$
$$= E(\mathbf{x} - E\mathbf{x}) \otimes (\mathbf{x} - E\mathbf{x}) = E\mathbf{x} \otimes \mathbf{x} - E\mathbf{x} \otimes E\mathbf{x}. \tag{4.31}$$

Definition 7. *A random vector* $\mathbf{x}$ *is called isotropic if, for any vector* $\mathbf{y} \in \mathbb{R}^n$, *the following equation holds:*

$$E < \mathbf{x}, \mathbf{y} >^2 = \|\mathbf{y}\|_2^2. \tag{4.32}$$

The notion of the isotropic random vector goes back to at least (Robertson, 1940); also, see (Rudelson, 1999; Kannan et al., 1997; Milman and Pajor, 1989) and (Vershynin, 2012).

Since $E < \mathbf{x}, \mathbf{y} >^2 = E(\mathbf{x}'\mathbf{y})^2 = <\Sigma\mathbf{y}, \mathbf{y}>$, the isotropic condition is equivalent to $<\Sigma\mathbf{y}, \mathbf{y}> = \|\mathbf{y}\|_2^2$. Next, since $<\Sigma\mathbf{x}, \mathbf{y}> = 1/4(<\Sigma(\mathbf{x}+\mathbf{y}), (\mathbf{x}+\mathbf{y})> - <\Sigma(\mathbf{x}-\mathbf{y}), (\mathbf{x}-\mathbf{y})>) = <\mathbf{x}, \mathbf{y}>$, the isotropic property is equivalent to $\Sigma = \mathbf{I}$.

Example 2. Gaussian. *The Gaussian random vector* $\mathbf{x}$ *with distribution* $N(0, \mathbf{I})$ *is isotropic. Indeed, it is true since the covariance matrix is* $\mathbf{I}$.

Bernoulli. *The n-dimensional Bernoulli random vector taking values in* $\{-1, 1\}^n$ *with equal probability is isotropic. Changing the sign of coordinates does not change expression in definition 4.32, hence* $\mathbf{x}$ *is isotropic.*

4.4 RIP for Matrices with Independent Bounded Rows and Matrices with Random Rows of Fourier Transform

The following theorem serves as a basis for rigorous explanation of the compressed sensing phenomenon.

Theorem 4.8. *(see (Candès et al., 2006a; Rudelson and Vershynin, 2008)) Let* $\mathbf{A} = (a_{ij})$ *be an* $n \times N$ *matrix which has independent normalized isotropic random vectors as columns. Suppose also that all entries are bounded,* $|a_{ij}| \leq K$. *Then for any* $\tau > 0$, *any* $N, k > 2$, *and with a probability at least* $1 - 5n^{-c\tau}$ *for*

$$N \geq C(K)\tau k \log^2(n) \, \log(C(K)\tau k \log^2(n)) \log^2(k), \tag{4.33}$$

$\mathbf{A}$ *satisfies k-RIP.*

When $k \geq \log n$, estimate 4.33 bounds N from below as

$$C_K'\tau \log(\tau)k \log^2(n) \log^3(k).$$

Our goal is to show that the mean of the RIP parameter

$$E = E\delta_k \tag{4.34}$$

is bounded around 1 in a controllable way. We will do that by expressing the estimate of E through E and resolving the inequality with respect to E.

Since vectors $\mathbf{A}_i$ are independent and isotropic, definition 4.32 is valid also for the vectors $\mathbf{x}$ with support in $T \subset \mathbb{Z}_N$ of size $|T| = d$. Hence, $\mathbf{A}_i|_T$ are independent and isotropic and $E\mathbf{A}_i|_T \otimes \mathbf{A}_i|_T = \mathbf{I}_d$.

Expression 1 may be presented as a tensor product of columns

$$\frac{1}{N}\mathbf{A}|_T^* \mathbf{A}|_T - \mathbf{I}_d = \frac{1}{N}\sum_{i=1}^{N} \mathbf{A}_i|_T \otimes \mathbf{A}_i|_T - \mathbf{I}_d = \frac{1}{N}\sum_{i=1}^{N} X_i, \qquad (4.35)$$

with $X_i = \mathbf{A}_i|_T \otimes \mathbf{A}_i|_T - \mathbf{I}_d$ being independent matrices with expectation 0.

Symmetrization step. We now estimate expression 1 through symmetrization of the X_i by random Bernoulli variable ϵ_i taking with equal probability values $\{-1, 1\}$, or

$$E \sup_{T\subset\mathbb{Z}_k} ||\sum_{i=1}^{N}(X_{iT} - EX_{iT})|| \le 2E \sup_{T\subset\mathbb{Z}_k} ||\sum_{i=1}^{N}\epsilon_i X_{iT}||. \qquad (4.36)$$

This form of the symmetrization inequality follows from the symmetrization inequality of section A.5 in Appendix. It states that the expected norm of the sum of centered variables is bounded from above by twice the expected norm of the sum of the symmetrized variables. Lemma A.10 implies estimate 4.36 as follows. We consider X_i as a vector column in $\mathbb{R}^n$ with the norm defined as

$$||X|| = ||X||_{\mathbb{R}^n} + \max_{T\subset\mathbb{Z}_k, |T|\le d}||X|_T||_{\mathbb{R}^n}. \qquad (4.37)$$

According to the note after definition 16 of vectors independence, vectors X_i are independent. Thus, we can apply symmetrization Lemma A.10.

By taking norm and then expectation in 4.35, we estimate E as

$$E = \frac{1}{N}E \sup_{T\subset\mathbb{Z}_N, |T|\le k} ||\sum_{j\in\mathbb{Z}_n} X_j|_T||.$$

Since $a_{j_1 j_2}$ is a function of original independent vectors $\mathbf{A}_j$, the vectors $X_j = \mathbf{A}_j \otimes \mathbf{A}_j$ are independent by note after definition 16. Hence, we can apply expression 4.36 in order to obtain

$$E \le \frac{2}{N}E \max_{T\subset\mathbb{Z}_N, |T|\le k} ||\sum_{j\in\mathbb{Z}_n}\epsilon_j \mathbf{A}_j|_T \otimes \mathbf{A}_j|_T||. \qquad (4.38)$$

We will now use the Uniform Rudelson Inequality (URI) (Rudelson, 1999, 2007; Vershynin, 2012).

Theorem 4.9. *Let $\mathbf{x}_1, ..., \mathbf{x}_l$ be vectors in $\mathbb{R}^m$, $l < m$, with entries bounded by constant $|x_{ji}| \le K$. For the $p \le m$ and independent Bernoulli random variables*

$\epsilon_1, ..., \epsilon_l$ holds

$$E \max_{T \subset \mathbb{Z}_m, |T| \leq p} || \sum_{i=1}^{l} \epsilon_i \mathbf{x}_i|_T \otimes \mathbf{x}_i|_T || \qquad (4.39)$$

$$\leq \phi(p, m, l) \sqrt{p} E \max_{T \subset \mathbb{Z}_m, |T| \leq p} || \sum_{i=1}^{l} \mathbf{x}_i|_T \otimes \mathbf{x}_i|_T ||^{\frac{1}{2}}, \qquad (4.40)$$

where $\phi(p, m, l) = C_K log(p) \sqrt{log(m) log(l)}$.

Estimating from above the right-hand side of inequality 4.38, using the result of Theorem 4.9, we obtain

$$E \leq \frac{\phi(k, n, N) \sqrt{k}}{N} E \max_{T \subset \mathbb{Z}_N, |T| \leq k} || \sum_{j \in \mathbb{Z}_N} \mathbf{x}_j|_T \otimes \mathbf{x}_j|_T ||^{\frac{1}{2}} \qquad (4.41)$$

$$= \frac{\phi(k, n, N) \sqrt{k}}{\sqrt{N}} E \max_{T \subset \mathbb{Z}_N, |T| \leq k} || \frac{1}{N} \mathbf{A}|_T \otimes \mathbf{A}|_T ||^{\frac{1}{2}}. \qquad (4.42)$$

Next, by using the triangle inequality, we get

$$E \max_{T \subset \mathbb{Z}_N, |T| \leq k} || \frac{1}{N} \mathbf{A}|_T \otimes \mathbf{A}|_T || \qquad (4.43)$$

$$\leq E \max_{T \subset \mathbb{Z}_N, |T| \leq k} || \frac{1}{N} \mathbf{A}|_T \otimes \mathbf{A}|_T - \mathbf{I}_k || + ||\mathbf{I}_k|| = E + 1. \qquad (4.44)$$

By substituting 4.44 into 4.42 we get

$$E \leq \phi(k, n, N) \sqrt{\frac{k}{N}} (E + 1)^{\frac{1}{2}}. \qquad (4.45)$$

The solution of inequality 4.45 for positive E is given by

$0 < E \leq \frac{1}{2}(a^2 + \sqrt{a^2(a^2 + 4)}) = b$, where $a = \phi(k, n, N) \sqrt{\frac{k}{N}}$. Since we are interested in the area where $a < 1$, we have $\sqrt{a^2 + 4} \leq a + 2$, and the right-hand side may be estimated above as $a(a + 1) \leq 2a$.

Figure 4.1 illustrates the estimates $b \leq 2a$ and $b \leq min\{\sqrt{2}a, 2a^2\}$.

By combining together the above inequalities, we conclude that, for the $0 < a \leq 1$,

$$E \leq 2a = 2C_K log(k) \sqrt{\frac{k \, log(n) log(N)}{N}}. \qquad (4.46)$$

By resolving 4.46 and by changing the constant C_K, we get

$$\frac{N}{log(N)} \geq C_K \delta^{-2} log^2(k) \, k \, log(n) \text{ implies } E \leq \delta. \qquad (4.47)$$

The following choice of N:

$$N \geq C_K \delta^{-2} log^2(k) \, k \, log(n) log(\delta^{-2} log^2(k) \, k \, log(n)) \qquad (4.48)$$

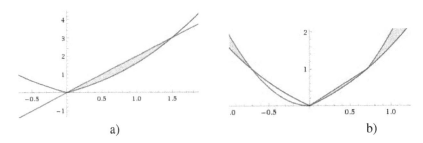

a)   b)

FIGURE 4.1: Areas of inequality: (a) area where $b \leq 2a$; (b) area where $b \leq max\{\sqrt{2}a, 2a^2\}$.

guaranties that, for $0 < \delta < 1$, we have $P(\{\delta_k > \delta^{\frac{1}{2}}\}) \leq \frac{E}{\delta^{\frac{1}{2}}}$, due to the simple estimate $P(\{\mathbf{x} > c\}) \leq E\mathbf{x}/c$. Thus, when $log(k)\sqrt{\frac{k \, log(n) log(N)}{N}}$ tends to 0, then k-RIP holds with probability of at least $1 - \delta^{\frac{1}{2}}$. This gives us at most logarithmic decay of deviation probability and, in principle, explains the phenomenon. In the rest of this section, we prove URI and show that RIP holds with probability $1 - n^{-c\tau}$.

4.4.1   Proof of URI

In this subsection, we will prove the URI Theorem 4.9, following the exposition presented in (Rudelson and Vershynin, 2008). The following estimate is due to Dudley; see section A.7 in Appendix.

Corollary 4.10. *(Dudley entropy inequality) Let $\mathcal{T}$ be a compact metric space. Let $V_t, t \in \mathcal{T}$ be a subgaussian process with $V_{t_0} = 0$. Then*

$$E \sup_{t \in \mathcal{T}} ||V_t|| \leq C \int_0^\infty \sqrt{log N(T, d, \epsilon)} d\epsilon. \tag{4.49}$$

Let B_2 be a unit ball in $\mathbb{R}^m$ with the Euclidean norm. We also consider B_1, the unit ball in l_1 norm in $\mathbb{R}^m$, and their restrictions B_1^T, B_2^T on coordinates $T \subset \mathbb{Z}_m$.

As a space $(\mathcal{T}, \delta)$ we choose $\bigcup_{T \subset \mathbb{Z}_m, |T| \leq k} B_2^T$. We consider metric

$$\delta(\mathbf{x}, \mathbf{y}) = \sup_{T \subset \mathbb{Z}_m, |T| \leq k} ||X_i|_T \otimes X_i|_T(\mathbf{x} \otimes \mathbf{x} - \mathbf{y} \otimes \mathbf{y})||_2 \tag{4.50}$$

$$= \sup_{T \subset \mathbb{Z}_q, |T| \leq k} \left[\sum_i^N (< X_i|_T, \mathbf{x} >^2 - < X_i|_T, \mathbf{y} >^2)^2 \right]^{\frac{1}{2}}. \tag{4.51}$$

We define a random process $V(\mathbf{x})$ as

$$V(\mathbf{x}) = \sum_i \epsilon_i X_i|_T \otimes X_i|_T \mathbf{x} \otimes \mathbf{x} \tag{4.52}$$

for $x \in \bigcup_{T \subset \mathbb{Z}_m, |T| \leq k} B_2^T$.

Note that $V(0) = 0$ and $V(x)$ is bounded, and hence it is subgaussian.

Denote the left-hand side of inequality 4.40 by E_1. Then, by using the definition of $X \otimes X$ and the Dudley inequality 4.49 we obtain

$$E_1 \leq E \max_{T \subset \mathbb{Z}_m, |T| \leq p; \mathbf{x} \in B_2^T} \left| \sum_{i=1}^l \epsilon_i < X_i|_T, \mathbf{x} >^2 \right| \tag{4.53}$$

$$\leq C \int_0^\infty log^{\frac{1}{2}} N(\bigcup_{T \subset \mathbb{Z}_m, |T| \leq p} B_2^T, \delta, u) du. \tag{4.54}$$

The metric δ on the right-hand side of 4.50 can be estimated as

$$\delta(\mathbf{x}, \mathbf{y}) \leq \left[\sum_{i=1}^l \left(< X_i|_T, \mathbf{x} >^2 + < X_i|_T, \mathbf{y} >^2 \right)^2 \right]^{\frac{1}{2}} \max_{i \leq l} | < X_i, \mathbf{x} - \mathbf{y} > |$$

$$\leq 2 \max_{T \subset \mathbb{Z}_m, |T| \leq p; \mathbf{x} \in B_2^T} \left[\sum_{i=1}^l < X_i|_T, \mathbf{x} >^2 \right]^{\frac{1}{2}} \max_{i \leq l} | < X_i, \mathbf{x} - \mathbf{y} > |$$

$$= 2R \max_{i \leq l} | < X_i, \mathbf{x} - \mathbf{y} > |, \tag{4.55}$$

where $R = \sup_{T \subset \mathbb{Z}_m, |T| \leq p} || \sum_{i=1}^l X_i|_T \otimes X_i|_T ||^{\frac{1}{2}}$.

Our goal now is to estimate the value of $N(\bigcup_{T \subset \mathbb{Z}_m, |T| \leq p} B_2^T, \delta, u)$. In order to do that, we replace B_2^T with B_1^T, which is easier to work with, and use metric δ on simpler $\max_{i \leq l} | < X_i, \mathbf{x} - \mathbf{y} > |$.

We will use the notation $D_q^{p,m} = \bigcup_{T \subset \mathbb{Z}_m, |T| \leq p} B_q^T$ for $q = 1, 2$. Let $||\mathbf{x}||_X = max_{i \leq l} | < X_i, \mathbf{x} > |$ and B_X be a unit ball in norm $||.||_X$. Note that $D_1^{p,m} \subset B_1^m$ and $D_1^{p,m} \subset K B_{1X}$, since $||\mathbf{x}||_X \leq K||\mathbf{x}||_1$. Similarly, since $||\mathbf{x}||_1 \leq \sqrt{p}||\mathbf{x}||_2$ (Cauchy-Schwarz inequality applied to $\mathbf{x}$ and vector with 1s only on the $supp(\mathbf{x})$), then $D_2^{p,m} \subset \sqrt{p} D_1^{p,m}$.

Thus, the right-hand side of 4.54 may be estimated from above as

$$C \int_0^\infty log^{\frac{1}{2}} N(\sqrt{p} \bigcup_{T \subset \mathbb{Z}_m, |T| \leq p} B_1^T, 2R||.||_X, u) du$$

$$\leq C \int_0^\infty log^{\frac{1}{2}} N(\sqrt{p} B_1^m, 2R||.||_X, u) du \tag{4.56}$$

$$\leq 2R\sqrt{p} C \int_0^\infty log^{\frac{1}{2}} N(B_1^m, ||.||_X, u) du,$$

since $N(\sqrt{p} B_1^m, 2R||.||_X, u) = N(B_1^m, ||.||_X, u/(2R\sqrt{p}))$ and the change of integration variable u to $u/(2R\sqrt{p})$ is applied.

To prove URI, it is enough to show that 4.56 does not exceed

$$C(K)\sqrt{p}\log(p)\sqrt{\log(m)\,\log(l)}R. \tag{4.57}$$

We partition the integral in two parts and estimate them separately. For the small u we use a volumetric estimate of Lemma 4.5 and Corollary 4.6. Recall that the corollary estimates size of the ϵ separated net. For the ϵ cover net we need to double the covering body, or take double norm. By taking into account that $B_1^T \subset KB_{1X}$, we estimate by $N(KB_{1X}, ||.||_X, u)$ as

$$
\begin{aligned}
N(B_1^T, ||.||_X, u) &\leq (1 + 2K/u)^p; \\
N(D_1^{p,m}, ||.||_X, u) &\leq d(m, p)(1 + 2K/u)^p,
\end{aligned}
\tag{4.58}
$$

where $d(m, p) = \sum_{i=1}^p mCi$ is a number of different B_1^T balls. We estimate mCi from above using the Stirling formula for $i! \sim \sqrt{2\pi i}(i/e)^i$, and possibly adjusting the constant as $Cn^i \cdot e^i/i^{(i+1/2)} = C^i/\sqrt{(i)}(m/i)^i$. The sum $d(m, p)$ does not exceed then $p(Cm/p)^p$. Then the $log^{\frac{1}{2}}$ of the right-hand side of expression 4.56 does not exceed

$$
C\left[\sqrt{\log(p)} + \sqrt{\log(m/p)} + \sqrt{\log(1 + 2K/u)}\right].
\tag{4.59}
$$

The following estimate of the covering number will be used for large u. We replace $\sqrt{p}D_1^{p,m}$ by $\sqrt{p}B_1^m$. Then

$$
N(B_1^m, ||.||_X, u) \leq (2m)^q,
\tag{4.60}
$$

where $q = C2K^2/u^2$. The proof is based on arguments due to Maurey; see (Carl, 1985; Rudelson and Vershynin, 2008).

We estimate that for the arbitrary vector y the distance from it to an appropriate equally weighted convex combination of q vectors with values ± 1 as the only nonzero coordinate is not more than u further in $||.||_X$ norm.

Indeed, fix vector $y \in B_1^m$ and consider random vector Z in $\mathbb{R}^m$, which takes the only nonzero value $sign(y_i)$ with probability $|y(i)|, i \in \mathbb{Z}_m$. Then $EZ = 0$.

Consider q identical copies of random variable Z, $Z_1, ..., Z_q$. Then, by using the symmetrization argument (see Lemma A.10),

$$
E||y - \frac{1}{q}\sum_{j\in\mathbb{Z}_q} Z_j||_X \leq \frac{2}{q}E||\sum_{j\in\mathbb{Z}_q} e_j Z_j||_X = C\frac{2}{q}E \max_{i\leq l}|\sum_{j\in\mathbb{Z}_q} <\epsilon_j Z_j, X_i >|.
\tag{4.61}
$$

Since $| < Z_j, X_i > | < K$, by the Cauchy inequality

$$
|\sum_{j\in\mathbb{Z}_q} <\epsilon_j Z_j, X_i >| \leq \sqrt{\sum_{j\in\mathbb{Z}_q} <\epsilon_j Z_j, X_i >^2} \leq K\sqrt{q}.
\tag{4.62}
$$

Corollary A.5 implies that the right-hand side of 4.61 does not exceed

$$
\frac{CK}{\sqrt{q}}.
\tag{4.63}
$$

Each Z_j takes $2m$ values, and $\frac{1}{q}\sum_{j\in\mathbb{Z}_q} Z_j$ takes possibly $(2m)^q$ values. For q chosen above, for arbitrary $y \in B_1^m$ holds $||y - \frac{1}{q}\sum_{j \in \mathbb{Z}_q} Z_j||_X \leq u$. By passing to the balls with double norm, we can restrict by choice of points in B_1^m.

The $log^{\frac{1}{2}}$ of the right-hand side of expression 4.60 does not exceed

$$\frac{CK \, \log^{\frac{1}{2}}(m)}{u}. \tag{4.64}$$

Now we return to the estimate 4.56. The integral 4.54 does not have any contribution from $u > K$, since $D_1^{p,m} \subset \sqrt{p}KB_{1X}$, and hence $N(\bigcup_{T \subset \mathbb{Z}_m, |T| \leq p} B_1^T, ||.||_X, u)$ is 1.

We partition integral 4.54 into two parts: integrals from 0 to $\mathbf{A} = \frac{1}{\sqrt{p}}$ and from $\mathbf{A}$ to K. The first part we estimate using 4.59, and the second part we estimate using 4.64.

For the first part, we estimate $\log^{\frac{1}{2}}$ as $\log$, since we suppose that $\mathbf{A} < 1/2$, and hence each addend is greater than 1, and integrate the estimate. Then the first part does not exceed (we again modified the constant without changing notation)

$$C\sqrt{p}\left[\mathbf{A}\sqrt{\log(p)} + \sqrt{\log(m/p)}\right] + \sqrt{p}(\mathbf{A}+K)\log((\mathbf{A}+K)(K))$$
$$\leq CRK \log(K)\sqrt{p}\log(p)\sqrt{\log(m)}. \tag{4.65}$$

The second part of the integral is estimated as (again, with a modified constant)

$$CRK\sqrt{p}\log(p)\sqrt{\log(m)}. \tag{4.66}$$

Thus, the whole integral 4.54 is estimated by $CRK\log(K)\sqrt{p}\log(p)\sqrt{\log(m)}$. This concludes the proof of the URI and the proof of 4.34.

4.4.2 Tail Bound for the Uniform Law of Large Numbers (ULLN)

In this section, we formulate and prove the tail estimate in the ULLN, and use it to finish the proof of Theorem 4.8. Again, our proof follows (Rudelson and Vershynin, 2008).

Consider random selectors, or n Bernoulli variables $\delta_1, ..., \delta_n$ taking value 1 with probability $\delta = k/n$. Define $\Omega = \{j \in \mathbb{Z}_n | \delta_j = 1\}$ and $p(\Omega) = k^{|\Omega|}(n-k)^{n-|\Omega|}/n^n$. Measure p defines probability spaces on all subsets of $\mathbb{Z}_n$. We say that Ω is a uniformly random set of size k since $E(|\Omega|) = k$.

Theorem 4.11. *Let $X_1, .., X_n$ be vectors in $\mathbb{R}^n$ with $|x_{ij}| < K$, for all i, j. Suppose that $\frac{1}{n}\sum_{i \in \mathbb{Z}_n} X_i \otimes X_i = \mathbf{I}$. Then the*

$$X = \sup_{T\mathbb{Z}_n, |T| \leq p} ||\sum_{i \in \mathbb{Z}_n} X_i|_T \otimes X_i|_T \mathbf{I}|| \tag{4.67}$$

is a random variable that satisfies for any $s > 1$,

$$p(X > Cs\epsilon) \leq 3e^{-C(K)s\epsilon k/r} + 2e^{-s^2}. \tag{4.68}$$

Proof. On the space $\mathcal{L}$ of linear operators from $\mathbb{R}^N \mapsto \mathbb{R}^N$ consider norm

$$||V||_{\mathcal{L}} = \sup_{T \in \mathbb{Z}_n, |T| \leq r} ||V|_T||, \qquad (4.69)$$

where $V|_T = P_T V P_T$, P_T is a projection on coordinates of T.

Let $\delta_1, ..., \delta_n$ be random selectors, i.e. i.i.d. Bernoulli variables with probability of having value 1 being k/n; $\delta'_1, ..., \delta'_n$ are their independent copies. Define random variables

$$x_i = \frac{1}{k}\delta_i \, X_i \otimes X_i - \frac{1}{n}\mathbf{I}_n, \qquad (4.70)$$

and y_i being symmetrization of x_i:

$$x_i = \frac{1}{k}(\delta_i - \delta'_i) \, X_i \otimes X_i. \qquad (4.71)$$

Define

$$X = ||\sum_{i \in \mathbb{Z}_n} x_i||_{\mathcal{L}}, \; Y = ||\sum_{i \in \mathbb{Z}_n} y_i||_{\mathcal{L}}. \qquad (4.72)$$

By Lemmas A.10, and A.35, A.36, the following estimates holds:

$$\begin{aligned} E(X) \leq E(Y) &\leq 2\,E(X) \\ p(X > 2E(X) + u) &\leq 2\,P(Y > u). \end{aligned} \qquad (4.73)$$

The norm $||X_i \otimes X_i||_{\mathcal{L}}$ is bounded as

$$||X_i \otimes X_i||_{\mathcal{L}} = \sup_{z \in \mathbb{R}^n, ||z||_2 = 1, |supp(z)| \leq r} |<x_i, z>|^2, \qquad (4.74)$$

and, hence

$$\begin{aligned} ||X_i \otimes X_i||_{\mathcal{L}} &\leq (||X_i||_{\infty} \sup_{||z||_2 = 1} ||z||_1)^2 \\ &\leq (||X_i||_{\infty}\sqrt{r} \sup_{||z||_2 = 1} ||z||_2)^2 \leq K^2 r. \end{aligned} \qquad (4.75)$$

The norm of y_i is estimated as

$$R = \max_{i \in \mathbb{Z}_n} ||y_i||_{calL} \leq \frac{2}{k}||X_i \otimes X_i||_{\mathcal{L}} \leq \frac{2K^2 r}{k} 2K^2 r. \qquad (4.76)$$

Now we apply to y_i the large deviation bound theorem; see Appendix, section A.8, Theorem A.13. Due to estimate 4.46, $E(X) < \delta$. Thus we have an estimate

$$p(X > (2 + 16q)\delta + 2Rl + t) \leq \frac{C^l}{q^l} + 2e^{-\frac{t^2}{512q\delta^2}}, \qquad (4.77)$$

for all natural $l \geq q, t > 0$. Set $q = floor(eC) + 1, t = \sqrt{512q}s\delta, l = floor(t/R)$. The condition $l \leq q$ holds due to 4.76 and choice of k. Thus

$$p(X > (2 + 16q + 3\sqrt{512q}s)\delta) \leq e^{-\frac{\sqrt{512q}s\delta k}{2K^2 r}} + 2e^{-s^2}. \qquad (4.78)$$

This estimates implies 4.69. □

In order to finish proof of the Theorem 4.8 we set $s = \frac{1}{2C\delta}$. Then $C(K)s\delta/r > 1/\delta^2$ by choice of N (or k). Hence

$$p(X > 1/2) \leq 5e^{-C/\delta^2}. \tag{4.79}$$

Since $\delta_r \leq X$, and $t = \mathbf{I}/\delta^2$, this finishes the proof of Theorem 4.8.

4.5 Summary and Bibliographical Notes

In this chapter, we consider RIP for the three main types of random matrices: matrices with subgaussian i.i.d. random entries, matrices with rows that are randomly selected from the Fourier (or cosine) transform matrix, and matrices with the rows selected randomly from an orthonormal matrix. The results considered in this section are due to (Donoho, 2006a; Candès et al., 2006a; Rudelson and Vershynin, 2008). Following (Baraniuk et al., 2008), we showed that RIP (and as corollary the Johnson-Lindenshtraus lemma) holds for the first type of matrices. Extension of these results to the matrices with sub-exponential entries can be found in (Adamczak et al., 2011), though with a slightly worse estimate. Note that the order of the estimate in 1 could not be improved, as it was shown in the phase-transition papers by (Donoho and Tanner, 2009) and also by (Foucart et al., 2010; Garnaev and Gluskin, 1984), since it is related to the lower bound of the Gelfand's width of l_1^N ball.

We also showed that RIP holds for the next two types of matrices, following (Rudelson and Vershynin, 2008). Further improvement of this estimate can be found in (Cheraghchi et al., 2013).

Note that the results presented in this chapter are uniform for any signal of fixed support. For non-uniform results that nevertheless imply recovery of the source, we refer to (Candès and Plan, 2011) and to several recent monographs (Eldar and Kutyniok, 2012; Foucart and Rauhut, 2013; Chafai et al., 2012).

Chapter 5

Algorithms for Sparse Recovery Problems

This chapter provides an overview of several common algorithms for sparse signal recovery, such as greedy approaches, active set methods (e.g., LARS algorithm), block-coordinate descent, iterative thresholding, and proximal methods. We focus on the noisy sparse recovery problems introduced before: the ultimate, and intractable, l_0-norm minimization:

$$(P_0^\epsilon): \quad \min_{\mathbf{x}} ||\mathbf{x}||_0 \text{ subject to } ||\mathbf{y} - \mathbf{Ax}||_2 \le \epsilon, \quad (5.1)$$

and its l_1-norm relaxation, also known as LASSO, or basis pursuit:

$$(P_1^\epsilon): \quad \min_{\mathbf{x}} ||\mathbf{x}||_1 \text{ subject to } ||\mathbf{y} - \mathbf{Ax}||_2 \le \epsilon, \quad (5.2)$$

frequently stated in its equivalent Lagrangian form:

$$(P_1^\lambda): \quad \min_{\mathbf{x}} \frac{1}{2}||\mathbf{y} - \mathbf{Ax}||_2^2 + \lambda||\mathbf{x}||_1. \quad (5.3)$$

Recall that $\mathbf{x}$ is an n-dimensional unknown sparse signal, which in a statistical setting corresponds to a vector of coefficients of a linear regression model, where each coefficient x_i signifies the amount of influence the i-th input, or predictor variable A_i, has on the output $\mathbf{y}$, an m-dimensional vector of observations of a *target* variable Y. $\mathbf{A}$ is an $m \times n$ design matrix, where the i-th column is an m-dimensional sample of a random variable A_i, i.e. a set of m independent and identically distributed, or i.i.d., observations.

71

Before we start discussing specific algorithms for the sparse recovery problems stated above, it is worth noting that problems (P_1^ϵ) and (P_1^λ), can be, of course, solved by general-purpose optimization techniques. For example, any convex unconstrained problem, such as (P_1^λ), can be handled by *subgradient descent*, an iterative method that, at each iteration, takes a step in the direction of the steepest decline of the objective. The approach is globally convergent; however, in practice, the convergence may be slow (Bach et al., 2012), and the solutions are typically not sparse. On the other hand, a more specific subclass of such methods, called proximal algorithms, is better suited for sparse problems and will be discussed later in this chapter.

Moreover, as it was already mentioned before, the problem (P_1^λ) can be formulated as a quadratic program, and thus general-purpose toolboxes, such as, for example, CVX[1], can be applied to it. This approach works well for relatively small problem sizes. However, as discussed in (Bach et al., 2012), generic quadratic programming does not scale well with increasing problem size, and thus it becomes necessary to exploit specific structure of the sparse recovery problem. Thus, the focus of this chapter is on specialized approaches for solving (P_1^λ) and (P_0^ϵ) problems, as well as some extensions of those problems to other types of objective functions and regularizers.

Also, before considering methods for solving the above problems in general, we would like to focus on the specific case of *orthogonal* design matrices. It turns out that in such case both l_0- and l_1-norm optimization problems decompose into independent *univariate* problems, and their optimal solutions can be easily found by very simple univariate *thresholding* procedures. This observation also provides an intuition for more general *iterative thresholding methods* described later in this chapter.

5.1   Univariate Thresholding is Optimal for Orthogonal Designs

An *orthogonal*, or *orthonormal*, matrix $\mathbf{A}$ is an $n \times n$ square matrix satisfying

$$\mathbf{A}^T\mathbf{A} = \mathbf{A}\mathbf{A}^T = \mathbf{I},$$

where $\mathbf{I}$ denotes the identity matrix, i.e. the matrix where the diagonal elements are all ones, and off-diagonal elements are all zeros. A linear transformation defined by an orthogonal matrix $\mathbf{A}$ has a nice property: it preserves the l_2-norm of a vector, i.e.

$$||\mathbf{A}\mathbf{x}||_2^2 = (\mathbf{A}\mathbf{x})^T(\mathbf{A}\mathbf{x}) = \mathbf{x}^T(\mathbf{A}^T\mathbf{A})\mathbf{x} = \mathbf{x}^T\mathbf{x} = ||\mathbf{x}||_2^2.$$

The same is clearly true for $\mathbf{A}^T$, and thus we get

$$||\mathbf{y} - \mathbf{A}\mathbf{x}||_2^2 = ||\mathbf{A}^T(\mathbf{y} - \mathbf{A}\mathbf{x})||_2^2 = ||\hat{\mathbf{x}} - \mathbf{x}||_2^2 = \sum_{i=1}^{n}(\hat{x}_i - x_i)^2,$$

---

[1] http://cvxr.com/cvx/.

where $\hat{\mathbf{x}} = \mathbf{A}^T\mathbf{y}$ corresponds to the ordinary least squares (OLS) solution when $\mathbf{A}$ is orthogonal, i.e.

$$\hat{\mathbf{x}} = \arg\min_{\mathbf{x}} ||\mathbf{y} - \mathbf{A}\mathbf{x}||^2.$$

As we show next, the above transformation of the sum-squared loss will greatly simplify both l_0- and l_1-norm optimizations problems.

5.1.1 l_0-norm Minimization

The problem (P_0^ϵ) can be now rewritten as

$$\min_{\mathbf{x}} ||\mathbf{x}||_0 \text{ subject to } \sum_{i=1}^{n} (\hat{x}_i - x_i)^2 \le \epsilon^2. \tag{5.4}$$

In other words, we are looking for the sparsest (i.e., smallest l_0-norm) solution $\mathbf{x}^*$ that is ϵ-close in l_2-sense to the OLS solution $\hat{\mathbf{x}} = \mathbf{A}^T\mathbf{y}$. It is easy to construct such solution by choosing k largest (in the absolute value) coordinates of $\hat{\mathbf{x}}$ and by setting the rest of the coordinates to zero, where k is the smallest number of such coordinates needed to get ϵ-close to $\hat{\mathbf{x}}$, i.e. to make the solution feasible. This can be also viewed as a *univariate hard-thresholding* of the OLS solution $\hat{\mathbf{x}}$, namely:

$$x_i^* = H(\hat{x}_i, \epsilon) = \begin{cases} \hat{x}_i & \text{if } |\hat{x}_i| \ge t(\epsilon) \\ 0 & \text{if } |\hat{x}_i| < t(\epsilon). \end{cases}$$

where $t(\epsilon)$ is a threshold value below the k-th largest, but above the $(k + 1)$-th largest value among $\{|\hat{x}_i|\}$. The univariate hard-thresholding operation, denoted here $H(x, \epsilon)$, is shown in Figure 5.1a.

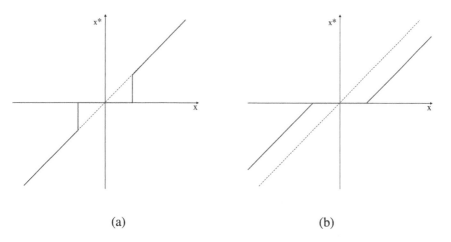

(a)   (b)

FIGURE 5.1: (a) Hard thresholding operator $x^* = H(x, \cdot)$; (b) soft-thresholding operator $x^* = S(x, \cdot)$.

5.1.2 l_1-norm Minimization

For an orthogonal **A**, the LASSO problem (P_1^λ) becomes

$$\min_{\mathbf{x}} \frac{1}{2} \sum_{i=1}^{n} (\hat{x}_i - x_i)^2 + \lambda \sum_{i=1}^{n} |x_i|, \tag{5.5}$$

which trivially decomposes into n independent, univariate optimization problems, one per each x_i variable, $i = 1, ..., n$:

$$\min_{x_i} \frac{1}{2} (\hat{x}_i - x_i)^2 + \lambda |x_i|. \tag{5.6}$$

When the objective function is convex and differentiable, it is easy to find its minimum by setting its derivative to zero. However, the objective functions in the above univariate minimization problems are convex but not differentiable at zero, because of $|x|$ being nondifferentiable at zero. Thus, we will use instead the notion of *subderivative*. Given a function $f(x)$, its *subdifferential* at x is defined as

$$\partial f(x) = \{z \in R^n | f(x') - f(x) \geq z(x' - x) \ for \ all \ x' \in R^n\},$$

where each z is called a *subderivative* and corresponds to the slope of a tangent line to $f(x)$ at x. Figure 5.2a shows the case of differentiable function where the subderivative is just the derivative at x (single tangent line). Figure 5.2a shows $f(x) = |x|$ as an example of a function nondifferentiable at $x = 0$, and its set of tangent lines

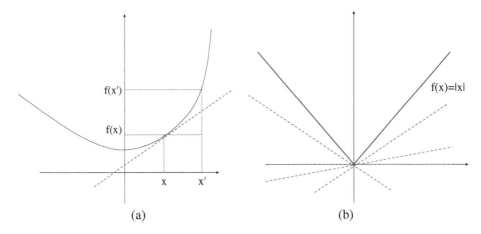

(a)   (b)

FIGURE 5.2: (a) For a function $f(x)$ differentiable at x, there is a unique tangent line at x, with the slope corresponding to the derivative; (b) a nondifferentiable function has multiple tangent lines, their slopes corresponding to subderivatives, for example, $f(x) = |x|$ at $x = 0$ has subdifferential $z \in [-1, 1]$.

corresponding to the subdifferential

$$\partial f(x) = \begin{cases} 1 & \text{if } x > 0 \\ [\text{-}1,1] & \text{if } x = 0 \\ -1 & \text{if } x < 0 \end{cases}.$$

Given a convex function $f(x)$, x^* is its global minimum if and only if $0 \in \partial f(x^*)$ (e.g., see (Boyd and Vandenberghe, 2004)). For the objective function

$$f(x) = \frac{1}{2}(x - \hat{x})^2 + \lambda|x|,$$

in each of the univariate LASSO problems given by eq. 5.6, the corresponding subdifferential is

$$\partial f(x) = x - \hat{x} + \lambda z, \tag{5.7}$$

and thus the condition $0 \in \partial f(x)$ implies

$$z = \begin{cases} 1 & \text{if } x > 0 \\ [\text{-}1,1] & \text{if } x = 0 \\ -1 & \text{if } x < 0. \end{cases}$$

This gives us the following global minimum solution for each univariate LASSO problem in eq. 5.6:

$$x_i^* = \begin{cases} \hat{x}_i - \lambda & \text{if } \hat{x}_i \geq \lambda \\ \hat{x}_i + \lambda & \text{if } \hat{x}_i \leq -\lambda \\ 0 & \text{if } |\hat{x}_i| < \lambda. \end{cases}$$

In other words,

$$x_i^* = S(\hat{x}, \lambda) = sign(\hat{x})(|\hat{x}| - \lambda)_+,$$

where the operator $S(\hat{x}, \lambda)$ transforming $\hat{x}_i$ into x_i^* is called the *soft thresholding* operator (shown in Figure 5.1b), as opposed to the *hard thresholding* operator $H(x, \epsilon)$ discussed earlier and illustrated in Figure 5.1a. In summary:

When the design matrix $\mathbf{A}$ is *orthogonal*, both l_0- and l_1-norm minimization problems decompose into a set of *independent univariate* problems given by eq. 5.4 and 5.6, respectively. These problems can be easily solved by first computing the OLS solution $\hat{\mathbf{x}} = \mathbf{A}^T \mathbf{y}$ and then applying thresholding operators to each coordinate: l_0-minimization (generally NP-hard combinatorial problem) is optimally solved by the univariate *hard thresholding*, while l_1-minimization (LASSO) is optimally solved by the univariate *soft thresholding*.

We will now go beyond the special case of orthogonal design matrices and consider several general-purpose optimization approaches for sparse recovery, starting with the ultimate l_0-norm minimization problem.

5.2   Algorithms for l_0-norm Minimization

As it was already mentioned, the ultimate sparse recovery problem, as defined by eq. 2.12, is NP-hard, and there are two categories of approximate techniques for solving it: those employing approximate methods, and those using convex approximations such as l_1-norm. In this section we focus on approximate optimization methods such as *greedy approaches* for solving the l_0-norm minimization combinatorial problem (P_0^ϵ):

$$(P_0^\epsilon): \quad \min_{\mathbf{x}} ||\mathbf{x}||_0 \text{ subject to } ||\mathbf{y} - \mathbf{Ax}||_2 \leq \epsilon, \quad (5.8)$$

which can be also written in an equivalent form as as

$$(P_0^k): \quad \min_{\mathbf{x}} ||\mathbf{y} - \mathbf{Ax}||_2 \text{ subject to } ||\mathbf{x}||_0 \leq k, \quad (5.9)$$

where the bound on the number of nonzero elements, k, is uniquely defined by the parameter ϵ in the original (P_0^ϵ) formulation. The latter problem is also known as the *best subset selection* problem, since it aims at finding a subset of k variables that yield the lowest quadratic loss in eq. 5.9, or, in other words, the best linear regression fit.

Instead of searching exhaustively for the best solution over all subsets of variables of size k, which is clearly intractable, a cheaper approximate alternative employed by greedy methods is to look for a single "best" variable at each iteration. A high-level greedy algorithmic scheme is outlined below.

Greedy Approach

1. Start with an empty *support set*, i.e. the set of nonzero ("active") variables, and zero vector as the current solution.

2. Select the best variable using some ranking criterion C_{rank}, and add the variable to the current support set.

3. Update the current solution, and recompute the current objective function, also called the *residual*.

4. If the current solution $\mathbf{x}$ satisfies a given stopping criterion C_{stop}, exit and return $\mathbf{x}$, otherwise go to step 2.

FIGURE 5.3: High-level scheme of greedy algorithms.

All greedy methods for sparse signal recovery that we consider here follow the above scheme. They differ, however, in specific ways the steps 2, 3, and 4 are implemented: namely, the choice of the variable ranking criterion C_{rank}, the way the current solution and the residual are updated, and the stopping criterion used. We will now discuss possible implementations of those steps, and the resulting greedy methods.

Step 2: Ranking Criteria for Best Variable Selection

Single-variable OLS fit. Let us consider the best subset selection problem in eq. 5.9, where $k = 1$. In other words, let us find best single-variable least-squares fit, i.e. a variable that yields the lowest sum-squared loss in 5.9. This problem can be easily solved in $O(N)$ time, by evaluating each column of the $M \times N$ matrix $\mathbf{A}$ that corresponds to a different variable. Namely, for each column $\mathbf{a}_i$, we will compute the corresponding sum-squared loss by minimizing the following univariate OLS objective:

$$L_i(x) = ||\mathbf{a}_i x - \mathbf{y}||_2^2 = ||\mathbf{y}||_2^2 - 2x\mathbf{a}_i^T \mathbf{y} + ||\mathbf{a}_i||_2^2 x^2. \tag{5.10}$$

Taking the derivative of the above function and setting it to zero gives

$$\frac{dL_i(x)}{dx} = -2\mathbf{a}_i^T \mathbf{y} + 2x||\mathbf{a}_i||_2^2 = 0,$$

which yields the optimal choice of the coefficient for the i-th variable:

$$\hat{x}_i = \arg\min_x L_i(x) = \mathbf{a}_i^T \mathbf{y}/||\mathbf{a}_i||_2^2. \tag{5.11}$$

The above single-variable OLS criterion is used to choose the next best variable in greedy methods such as matching pursuit (MP), also known as forward stagewise regression in statistics, and in orthogonal matching pursuit (OMP); these algorithms will be discussed in more detail later in this chapter.

Full OLS fit. Another possible evaluation criterion for selecting the best next variable is to compute the overall improvement in the objective function in eq. 5.9 that can be gained due to adding a candidate variable to the current support. This requires solving the full OLS regression problem, where the set of variables is restricted to the current support set, plus the candidate variable. Let $S = \{i_1, ..., i_k\}$ be the current support set of size k, where i_j is the index of the variable (column of $\mathbf{A}$) added at the j-th iteration, let i be the index of the column/variable that is being evaluated, and let $\mathbf{A}|_S$ be the restriction of the matrix $\mathbf{A}$ on a subset of columns indexed by S. Then the full-OLS approach is to compute the current solution estimate as

$$\hat{\mathbf{x}} = \arg\min_{\mathbf{x}} ||\mathbf{y} - \mathbf{A}|_{S\cup\{i\}} \mathbf{x}||_2^2. \tag{5.12}$$

Then the current estimate of $\mathbf{y}$ is given by

$$\hat{\mathbf{y}} = \mathbf{A}|_{S\cup\{i\}} \hat{\mathbf{x}}, \tag{5.13}$$

and the new *residual*, i.e., the remaining part of $\mathbf{y}$ not yet "explained" by the current solution, is given by

$$\mathbf{r} = \mathbf{y} - \hat{\mathbf{y}}. \tag{5.14}$$

Thus, according to the full-OLS criterion, the best next variable is the one that results into the lowest residual, i.e. the smallest sum-squared loss in eq. 5.12. Note that, while the full-OLS approach is computationally more expensive than the single-variable fit, it often leads to more accurate approximations of the exact sparse solution, as shown, for example, in (Elad, 2010); also, see chapter 3 of (Elad, 2010) for an efficient implementation of this step. The full-OLS fit approach to variable selection is used in the Least-Squares OMP (LS-OMP), as discussed in (Elad, 2010), and is also widely known as *forward stepwise regression* in statistics (Hastie et al., 2009; Weisberg, 1980).

Comments. It will be useful to keep in mind the following geometric interpretation of the full-OLS fit: estimate $\hat{\mathbf{y}}$ can be viewed as the *orthogonal projection* of the vector $\mathbf{y}$ on the column-space of $\mathbf{A}|_{S \cup \{i\}}$, and thus the corresponding residual in eq. 5.14 is *orthogonal* to that space.

Also, note that, if the columns of $\mathbf{A}$ are normalized to have the unit l_2-norm, i.e. $||\mathbf{a}_i||_2 = 1$, then the estimate $\hat{x}_i = \mathbf{a}_i^T \mathbf{y}$. And, if on top of that, both $\mathbf{y}$ and the columns of $\mathbf{A}$ are centered to have zero means, then the correlation between the observed values $\mathbf{a}_i$ of the i-th predictor variable and the observed values $\mathbf{y}$ of the predicted variable, can be written as

$$corr(\mathbf{y}, \mathbf{a}_i) = \mathbf{a}_i^T \mathbf{y}/||\mathbf{y}||_2.$$

Thus, *for normalized and centered data, the best single-variable fit is achieved by the column $\mathbf{a}_i$ most-correlated with* $\mathbf{y}$. Moreover, since

$$\cos \alpha = \frac{\mathbf{a}_i^T \mathbf{y}}{||\mathbf{a}_i||_2 ||\mathbf{y}||_2},$$

where α is the angle between $\mathbf{a}_i$ and $\mathbf{y}$, *the two column-vectors $\mathbf{a}_i$ and $\mathbf{a}_j$ that are equally correlated with $\mathbf{y}$ also have an equal angle with it.*

Step 3: Updating the Current Solution and the Residual

Once the next best variable i is selected, it is added to the current support. However, there are several possible options when it comes to updating the current solution and the residual. If a single-variable fit was used at step 2, then the coefficients of the previously selected variables are all left intact, and only the coefficient of the currently selected variable i is updated, followed by the update of the current residual, as follows:

$$x_i = x_i + \hat{x}_i, \tag{5.15}$$

$$\mathbf{r} = \mathbf{r} - \mathbf{a}_i^T \hat{x}_i. \tag{5.16}$$

A more computationally involved update that often leads to a better performance (Elad, 2010) is to actually recompute the current solution, over the new support $S \cup \{i\}$, by solving the OLS problem in eq. 5.12, which also produces the new residual (see eq. 5.14). Note that the residual obtained via such update is *orthogonal* to the column-space of $\mathbf{A}$ restricted to the current solution support, and thus none of the variables in the current support will be considered as potential candidates in the future iterations of the algorithm, while this is not necessarily true when using a simple update in eq. 5.16. The update orthogonalizing the residual is used in the greedy method known as orthogonal matching pursuit (OMP).

Step 4: Stopping Condition

A commonly used stopping condition is achieving some sufficiently small error on the objective function; the error threshold is given as an input to the algorithm. For example, setting the error threshold to ϵ in the formulation (P_0^ϵ) guarantees that the solution will satisfy the constraint $||y - Ax||_2 \leq \epsilon$. Another possible criterion, used, for example, in forwards stagewise regression formulation (Hastie et al., 2009), is to stop the algorithm when there are no more predictors (columns of A) that are still correlated with the current residual, so that no more improvement to the objective function is possible.

5.2.1 An Overview of Greedy Methods

Greedy algorithms for sparse signal recovery have a long history in both signal processing and statistics fields, and same algorithms sometimes appear under different names in different research communities. Here we present several most commonly-used methods, although multiple extensions and modifications exist along the lines of the general greedy scheme present above. From now on, we will assume that the vector y is centered to have zero mean, and that the columns of the matrix A are normalized to the unit l_2-norm and centered to have zero mean, as it is commonly assumed in statistical literature. Recall that under such assumptions the best single-variable fit to the current residual r is achieved by the column a_i most-correlated with the residual.

The following algorithm, shown in Figure 5.4, is known as *Matching Pursuit (MP)* (Mallat and Zhang, 1993; Elad, 2010) in the signal processing community, and is essentially equivalent to the *forward stagewise regression* (Hastie et al., 2009) in statistics, although the stopping criteria for those two algorithms may be stated differently: MP formulations typically use a threshold on the sum-squared error (Elad, 2010), while forward stagewise iterates until there are no more predictors (columns of A) correlated with the current residual (Hastie et al., 2009). The MP algorithm is the simplest greedy method out of those considered herein; it uses simple single-variable OLS fit to find the next best variable, and a simple update of the solution and the residual, which only involves a change to the coefficient of the selected variable. However, as mentioned before, such update does not orthogonalize the residual with respect to the variables in the current support, and thus the same variable may be selected again in the future. Thus, though each iteration of the algorithm is simple, MP, or forward stagewise, may require a potentially large number of iterations to converge.

A modification of the forward stagewise regression algorithm, known as the *incremental forward stagewise regression* (Hastie et al., 2009), makes smaller

---

Matching Pursuit (MP)
Input: A, y, ϵ

1. **Initialize:** $k = 0$, $\mathbf{x}^k = 0$, $S^k = supp(\mathbf{x}^0) = \emptyset$, $\mathbf{r}^k = \mathbf{y} - \mathbf{A}\mathbf{x}^0 = \mathbf{y}$

2. **Select next variable:**
   $k = k + 1$ // next iteration
   find the predictor (column in $\mathbf{A}$) most correlated with the residual $\mathbf{r}^{k-1}$:
   $i^k = \arg\max_i \mathbf{a}_i^T \mathbf{r}^{k-1}$
   $\hat{x}_{i^k} = \max_i \mathbf{a}_i^T \mathbf{r}^{k-1}$

3. **Update:**
   $S^k \leftarrow S^{k-1} \cup \{i^k\}$ // update the support
   $\mathbf{x}^k = \mathbf{x}^{k-1}$, $x_i^k = x_i^k + \hat{x}_{i^k}$ // update the solution
   $\mathbf{r}^k = \mathbf{r}^{k-1} - \mathbf{a}_i^T \hat{x}_{i^k}$ // update the residual

4. **If** the current residual $\mathbf{r}^k$ is uncorrelated with all $\mathbf{a}_i$, $1 \le i \le n$, or
   if $||\mathbf{r}^k||_2 \le \epsilon$, **then** return $\mathbf{x}^k$,
   otherwise go to step 2 (next iteration of the algorithm).

FIGURE 5.4: Matching Pursuit (MP), or forward stagewise regression.

(incremental) updates at step 3 of the algorithm shown in Figure 5.4, i.e. $x_i^k = x_i^k + \delta \cdot \hat{x}_{i^k}$, where the parameter δ controls the step size. Incremental forward stagewise regression with $\delta \to 0$ turns out to be closely related to the Lasso problem and the LARS algorithm for solving it, as discussed in the next section.

We now present the next algorithm in the greedy family, known as the *Orthogonal Matching Pursuit (OMP)*, which was introduced shortly after MP by both (Pati et al., 1993) and (Mallat et al., 1994). As shown in Figure 5.5, it differs from MP only in the way the solution and the residual are updated (step 3). As it was mentioned above when discussing different implementations of the step 3, OMP will recompute the coefficients of *all* variables in the current support, by solving the full OLS problem over the support augmented with the new variable. As the result of this operation, the residual becomes orthogonal to the support variables, hence the word "orthogonal" in the name of the algorithm. Again, as mentioned above, the OMP update step is more computationally expensive than the MP update, but, due to orthogonalization, it will only consider each variable once, typically resulting into a smaller number of iterations. Also, OMP often obtains more accurate sparse solutions than MP.

Finally, progressing from simple to more sophisticated greedy approaches, we consider the so-called *Least-Squares OMP (LS-OMP)* algorithm presented in (Elad, 2010), which is also widely known in statistical literature as *forward stepwise regression* (Hastie et al., 2009); see Figure 5.6 for details. This approach is sometimes confused with OMP (e.g., see (Blumensath and Davies, 2007) for a detailed discussion and historical remarks on both algorithms), and thus it is important to clarify

---

Orthogonal Matching Pursuit (OMP)
Input: A, y, ϵ

1. **Initialize:** $k = 0$, $x^k = 0$, $S^k = supp(x^0) = \emptyset$, $r^k = y - Ax^0 = y$

2. **Select next variable:**
   $k = k + 1$ // next iteration
   find the predictor (column in A) most correlated with the residual r^{k-1}:
   $i^k = \arg\max_i a_i^T r^{k-1}$
   $\hat{x}_{i^k} = \max_i a_i^T r^{k-1}$

3. **Update:**
   $S^k \leftarrow S^{k-1} \cup \{i^k\}$ // update the support
   $x^k = \arg\min_x \|y - A|_{S^k} x\|_2^2$ // full-OLS fit on the updated support
   $r^k = \min_x \|y - A|_{S^k} x\|_2^2$ // update the residual

4. **If** the current residual r^k is uncorrelated with all a_i, $1 \leq i \leq n$, or
   if $\|r^k\|_2 \leq \epsilon$, **then** return x^k,
   otherwise go to step 2 (next iteration of the algorithm).

FIGURE 5.5: Orthogonal Matching Pursuit (OMP).

the distinction between OMP and forward stepwise regression. The key difference between the two methods is in the variable-selection criterion used in step 2: while OMP, similarly to MP, finds the predictor variable most correlated with the current residual (i.e., performs the single-variable OLS fit), LS-OMP, or forwards stepwise regression, searches for a predictor that best improves the *overall fit*, i.e. solves the full OLS problem on the current support plus the candidate variable. Though this step is more computationally expensive than the single-variable fit, efficient implementations are available that speed it up–see, for example, (Elad, 2010; Hastie et al., 2009). As a result, all entries in the current solution are updated, so the step 3 of LS-OMP (i.e., updating the solution and the residual) coincides with the step 3 of OMP.

To summarize, we discussed herein three commonly used greedy methods for the best subset selection, or sparse recovery, problem: Matching Pursuit (MP), also known as forward stagewise regression; Orthogonal Matching Pursuit (OMP), and the Least-Squares OMP (LS-OMP), equivalent to forward stepwise regression. However, as it was already mentioned above, there are multiple extensions and improvements to the basic greedy schemes, including Stagewise OMP (StOMP) (Donoho et al., 2012), compressive sampling matching pursuit (CoSaMP) (Needell and Tropp, 2008), regularized OMP (ROMP) (Needell and Vershynin, 2009), subspace pursuit (SP) (Dai and Milenkovic, 2009), sparsity adaptive matching pursuit (SAMP) (Do et al., 2008), and several others.

Least-Squares Orthogonal Matching Pursuit (LS-OMP)
Input: A, y, ϵ

1. **Initialize:** $k = 0$, $\mathbf{x}^k = 0$, $S^k = supp(\mathbf{x}^0) = \emptyset$, $\mathbf{r}^k = \mathbf{y} - \mathbf{A}\mathbf{x}^0 = \mathbf{y}$

2. **Select next variable:**
   $k = k + 1$ // next iteration
   find the predictor (column in $\mathbf{A}$) that most improves the full-OLS fit over S^{k-1}:
   $i^k = \arg\min_i \min_{\mathbf{x}} ||\mathbf{y} - \mathbf{A}|_{S \cup \{i\}}\mathbf{x}||_2^2.$
   $\hat{x}_{i^k} = \arg\min_{\mathbf{x}} ||\mathbf{y} - \mathbf{A}|_{S \cup \{i^k\}}\mathbf{x}||_2^2$

3. **Update:**
   $S^k \leftarrow S^{k-1} \cup \{i^k\}$ // update the support
   $\mathbf{x}^k = \arg\min_{\mathbf{x}} ||\mathbf{y} - \mathbf{A}|_{S^k}\mathbf{x}||_2^2$ // full-OLS fit on the updated support
   $\mathbf{r}^k = \min_{\mathbf{x}} ||\mathbf{y} - \mathbf{A}|_{S^k}\mathbf{x}||_2^2$ // update the residual

4. **If** the current residual $\mathbf{r}^k$ is uncorrelated with all $\mathbf{a}_i$, $1 \leq i \leq n$, or
   if $||\mathbf{r}^k||_2 \leq \epsilon$, **then** return $\mathbf{x}^k$,
   otherwise go to step 2 (next iteration of the algorithm).

FIGURE 5.6: Least-Squares Orthogonal Matching Pursuit (LS-OMP), or forward stepwise regression.

5.3   Algorithms for l_1-norm Minimization (LASSO)

As it was already mentioned, an alternative approach to solving the intractable l_0-norm minimization problem P_0^ϵ is to replace the l_0-norm by a convex function which still enforces the solution sparsity, such as, for example, the l_1-norm, that gives us the (P_1^ϵ) problem, or its equivalent and commonly used Lagrangian form – the LASSO problem (P_1^λ). There is a vast amount of literature on different methods for solving this problem, and multiple software packages available online (see, for example, the Compressed Sensing repository at http://dsp.rice.edu/cs). Herein, we briefly review several most commonly used algorithms for solving LASSO, starting with the LARS method of (Efron et al., 2004).

5.3.1   Least Angle Regression for LASSO (LARS)

Least Angle Regression (LAR) proposed by (Efron et al., 2004) is a modification of the incremental forward stagewise procedure. As shown in (Efron et al., 2004), after a minor modification, LAR becomes equivalent to the LASSO, and another simple modification of LAR makes it equivalent to *infinitesimal forward stagewise*, i.e. incremental forward stagewise with the step size $\epsilon \to 0$, denoted FS_0, as discussed in (Hastie et al., 2009). Frequently, LAR is also called LARS, where "S" is added to denote the close relationship to both Stagewise and LaSSo. One of the most attractive properties of LARS is its computational efficiency: at the cost of a single OLS fit, LARS finds the entire *regularization path*, i.e. the sequence of all LASSO solutions corresponding to varying regularization parameter $\lambda > 0$:

$$\hat{\mathbf{x}}(\lambda) = \arg\min_{\mathbf{x}} f(\mathbf{x}, \lambda) = \arg\min_{\mathbf{x}} \frac{1}{2}||\mathbf{y} - \mathbf{A}\mathbf{x}||_2^2 + \lambda||\mathbf{x}||_1.$$

The efficiency of LARS results from the fact that the LASSO path is *piecewise linear*, as it was shown earlier in chapter 2; indeed, at each step corresponding to a change in the path direction, LARS must only evaluate the new direction and the step length, thus avoiding inefficient incremental steps. The piecewise linearity of the LASSO solution path was also used by a similar to LARS *homotopy method* proposed earlier in (Osborne et al., 2000a). Note that an efficient way of computing the full regularization path simplifies the cross-validation procedure of selecting the regularization parameter that yields the best-predicting solution on a set-aside (cross-validation) data.

LARS can be viewed, on one hand, as a more cautious version of the forward stepwise regression, and, on the other hand, as a more efficient version of the incremental forward stagewise, since it makes smaller steps than the former, but larger steps than the latter. More specifically, LARS starts with an empty set of predictors and selects the one having the largest absolute correlation with the response; however, unlike the overly greedy forward stepwise, LARS proceeds along the selected direction only until another predictor becomes equally correlated (in the absolute sense) with the current residual. Then, LARS chooses a new direction, which is equiangular between the two predictors[2], and continues moving along this direction until some third predictor enters the "most correlated" set, also called the *active set*. LARS chooses the new direction equiangular between the three active predictors, and continues further, until the desired number of predictors, that can be specified as an input to the algorithm, is included in the solution, or until the full path is obtained, which takes $\min(m - 1, n)$ steps. A high-level scheme of LARS is presented in Figure 5.7. As before, it is assumed that all $\mathbf{A}_i$ and $\mathbf{y}$ are centered to have zero means, and that all $\mathbf{A}_i$ are normalized to have to unit norms. For more details on LARS, see (Efron et al., 2004). Multiple publicly available implementations of this algorithm can be found online. See, for example, Splus and R implementations of LARS (Efron and Hastie, 2004), a Matlab implementation by (Sjöstrand, 2005), and several others.

LASSO modification. As it was shown by (Efron et al., 2004), LARS becomes equivalent to LASSO, i.e. it finds the path of *optimal* solutions to the LASSO problem, if the following minor modification is added: *If a nonzero coefficient becomes zero, the corresponding variable is removed from the active set, and the current joint least-squares direction for LARS is recomputed over the remaining variables.* For

---

[2]Recall that two column-vectors $\mathbf{a}_i$ and $\mathbf{a}_j$ that are equally correlated with $\mathbf{y}$ also have an equal angle with it, as discussed before.

Least Angle Regression (LAR)
Input: $m \times n$ matrix $\mathbf{A}, \mathbf{y}$

1. **Initialize:**
   $\mathbf{x} = 0, S = supp(\mathbf{x}) = \emptyset, \mathbf{r} = \mathbf{y} - \mathbf{Ax} = \mathbf{y}$

2. **Select first variable:**
   find the predictor (column in $\mathbf{A}$) most correlated with the residual $\mathbf{r}$:
   $i = \arg\max_i \mathbf{a}_i^T \mathbf{r}$
   $\hat{x}_i = \max_i \mathbf{a}_i^T \mathbf{r}$
   $S \leftarrow S \cup \{i\}$ // update the support

3. Move the coefficient x_i from 0 towards its least-squares coefficient $\hat{x}_{i^k}$, updating the residual $\mathbf{r}$ along the way, until some other predictor $\mathbf{a}_j$ has as much correlation with the current residual as does $\mathbf{a}_i$; then add it to the support: $S \leftarrow S \cup \{j\}$.

4. Move x_i and x_j in the direction defined by their joint least-squares coefficient:
$$\delta_k = (\mathbf{A}_{S^k}^T \mathbf{A}_{S^k})^{-1} \mathbf{A}_{S^k}^T \mathbf{r}$$
   of the current residual on the current support set S, until some other predictor $\mathbf{a}_k$ has as much correlation with the current residual; then add it to the support: $S \leftarrow S \cup \{k\}$.

5. Continue adding predictors for $\min(m-1, n)$ steps, until full OLS solution is obtained. If $n < m$, all predictors are now in the model.

FIGURE 5.7: Least Angle Regression (LAR).

example, Figure 5.8 shows an example comparing the path of LARS before and after implementing the above modification: a nonzero coefficient of one of the variables actually reaches zero, and the variable has to be deleted from the active set.

To summarize, LARS is a stagewise procedure closely related to greedy approaches such as OMP, forward stepwise (LS-OMP), and forward stagewise (MP), but, unlike the greedy methods, LARS (after a minor modification) is able to produce the exact solution to the LASSO problem (P_1^λ) in eq. 2.14.

- LARS produces the same solution path as LASSO, if the coefficients do not cross zero (otherwise, LASSO modification is applied to achieve the equivalence).

- LARS produces same solution path as FS_0 (infinitesimal forward stagewise) if the coefficients are monotone (otherwise, another simple FS_0 modification, discussed in (Hastie et al., 2009), can be applied to achieve the equivalence between the methods).

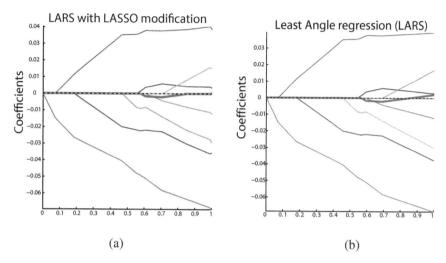

(a)   (b)

FIGURE 5.8: Comparing regularization path of LARS (a) before and (b) after adding the LASSO modification, on fMRI dataset collected during the pain perception analysis experiment in (Rish et al., 2010), where the pain level reported by a subject was predicted from the subject's fMRI data. The x-axis represents the l_1-norm of the sparse solution obtained on the k-th iteration of LARS, normalized by the maximum such l_1-norm across all solutions. For illustration purposes only, the high-dimensional fMRI dataset was reduced to a smaller number of voxels ($n = 4000$ predictors), and only $m = 9$ (out of 120) samples were used, in order to avoid clutter when plotting the regularization path. Herein, LARS selected $\min(m - 1, n) = 8$ variables and stopped.

- LARS is very efficient: at the cost of solving OLS, LARS produces the *whole regularization (solution) path*, i.e. a sequence of solutions for the varying values of the regularization parameter λ, from very large (corresponding to an empty solution), to zero (corresponding to OLS solution).

As we noted above, such efficient computation of the LASSO regularization path is possible because the path is piecewise linear. As it turns out, this nice property also holds for a more general class of regularized problems

$$\hat{\mathbf{x}}(\lambda) = \arg\min_{\mathbf{x}}[L(\mathbf{x}) + \lambda R(\mathbf{x})],$$

with a convex loss L and a regularizer R, where

$$L(\mathbf{x}) = \sum_{i=1}^{m} Loss(y_i, \sum_{j}^{n} a_{ij}x_j).$$

Namely, as shown by (Rosset and Zhu, 2007), the solution path is piecewise linear if $L(\mathbf{x})$ is quadratic or piecewise quadratic, and $R(\mathbf{x})$ is piecewise linear, as it is in case of l_1-norm.

5.3.2   Coordinate Descent

Another popular approach to solving the LASSO problem, as well as its generalizations to other loss functions and regularizers discussed later in this book, is *coordinate descent*. As it name suggests, this commonly used optimization technique fixes all variables (coordinates) except for one, and performs univariate optimization with respect to that single remaining variable. The procedure keeps iterating over all variables in some cyclic order, until convergence[3]. Coordinate descent is an efficient procedure that is competitive with, and often better than, the LARS algorithm on the LASSO problem (Hastie et al., 2009), due to an efficient implementation of the univariate optimization step, as we discuss below. However, unlike LARS, coordinate descent is not a path-generating method, and can only approximate the solution path by computing the solutions using warm-restart on a *grid* of λ values.

As before, we assume that all $\mathbf{A}_i$ and $\mathbf{y}$ are centered to have zero means, and that all $\mathbf{a}_i$ are normalized to have the unit norm. Let us denote the current estimate of x_i by $\hat{x}_i(\lambda)$, where λ is the regularization parameter of the LASSO problem in eq. 2.14. Assuming that all coefficients, except for the i-th, are fixed, we can rewrite the objective function in eq. 2.14 as a function of a single variable x_i:

$$L(x_i, \lambda) = \frac{1}{2}\sum_{j=1}^{m}([y_j - \sum_{k \neq i} a_{jk}\hat{x}_k(\lambda)] - a_{ji}x_i)^2 + \lambda \sum_{k \neq i}|\hat{x}_k(\lambda)| + \lambda|x_i| \quad (5.17)$$

$$= \frac{1}{2}||\mathbf{r} - \mathbf{a}_i^T x_i||_2^2 + \lambda|x_i| + const, \quad (5.18)$$

where the current residual vector $\mathbf{r}$ has elements $r_j = y_j - \sum_{k \neq i} a_{jk}\hat{x}_k(\lambda)$, and thus the above objective corresponds to a univariate LASSO problem. Similarly to the derivation of the LASSO solution in case of the orthogonal design matrix, discussed earlier in this chapter, we compute the subdifferential of $L(x_i, \hat{\mathbf{x}}(\lambda))$ as follows:

$$\partial L(x_i, \lambda) = \sum_{j=1}^{m} a_{ji}(a_{ji}x - r_j) + \lambda\partial|x| =$$

$$= \mathbf{a}_i^T \mathbf{a}_i x - \mathbf{a}_i^T \mathbf{r} + \lambda\partial|x| = x - \mathbf{a}_i^T \mathbf{r} + \lambda\partial|x|,$$

where $\mathbf{a}_i^T \mathbf{a}_i = 1$ since columns of $\mathbf{A}$ are normalized to unit l_2-norm. Similarly to eq. 5.7, the condition $0 \in \partial L(x, \lambda)$ implies that the univariate LASSO solution x will be given by the soft thresholding operator applied to the univariate OLS solution $\mathbf{a}_i^T \mathbf{r}$, and thus

$$\hat{x}_i = S(\mathbf{a}_i^T \mathbf{r}, \lambda) = sign(\mathbf{a}_i^T \mathbf{r})(|\mathbf{a}_i^T \mathbf{r}| - \lambda)_+ \quad (5.19)$$

gives the update rule for the i-th coefficient x_i. Coordinate descent continues iterating over the variables in some pre-specified cyclic order, until convergence, as noted above, and is generally quite fast due to the efficiency of computations at each step given by eq. 5.19 above. A scheme of the algorithm is outlined in Figure 5.9.

---

[3]See (Bach et al., 2012) and references therein on convergence of coordinate descent with non-smooth objective functions.

Coordinate Descent (CD)
Input: $m \times n$ matrix $\mathbf{A}$, $\mathbf{y}$, λ, ϵ

1. **Initialize:** $k = 0$, $\mathbf{x} = 0$, $\mathbf{r} = \mathbf{y}$
   Cycle over all variables until convergence:

2. $k = k + 1$, $i = k \bmod (n + 1)$

3. Compute partial residual $\mathbf{r}$ where:

$$r_j = y_j - \sum_{k \neq i} a_{jk} x_k$$

4. Compute univariate OLS solution:

$$x = \mathbf{a}_i^T \mathbf{r}$$

5. Update x_i (univariate LASSO solution):

$$x_i = S(x, \lambda) = sign(x)(|x| - \lambda)_+$$

6. **If** the (full) residual $\mathbf{r} - \mathbf{a}_i x_i < \epsilon$, return $\mathbf{x}$,
   otherwise go to step 2 (next iteration of the algorithm).

FIGURE 5.9: Coordinate descent (CD) for LASSO.

Note that there are multiple extensions of the above coordinate descent algorithm to other loss functions, beyond the sum-squared loss in LASSO, e.g., for sparse logistic regression (Tseng and Yun, 2009) and for sparse Gaussian Markov Network models (Banerjee et al., 2006; Friedman et al., 2007b), discussed in chapter 8, as well as extensions beyond the basic l_1-norm regularization, e.g., to structured sparsity via group regularizers such as l_1/l_q (Yuan and Lin, 2006), discussed later in this book.

5.3.3 Proximal Methods

In this section, we introduce *proximal methods*, also known as *forward-backward splitting* algorithms – a general class of convex optimization techniques that include as special cases many well-known algorithms, such as iterative thresholding, subgradient and projected gradient methods, as well as alternating projections, among several others; for more details see (Combettes and Pesquet, 2011; Bach et al., 2012) and references therein. Proximal methods have a long history in optimization; relatively recently, i.e. during the past decade, they were introduced in signal processing and sparse optimization, and quickly gained popularity there.

5.3.3.1 Formulation

Let us consider the following general problem that includes as special cases multiple sparse-recovery problems, such as the basic LASSO and its extensions beyond the l_1-norm regularizer and beyond the quadratic loss, discussed in other chapters of this book:

$$(P): \quad \min_{\mathbf{x}} f(\mathbf{x}) + g(\mathbf{x}), \qquad (5.20)$$

where

- $f : R^n \to R$ is a smooth convex function, which is continuously differentiable with Lipshitz continuous gradient, i.e.

$$||\nabla f(\mathbf{x}) - \nabla f(\mathbf{y})|| \le L(f)||\mathbf{x} - \mathbf{y}||, \quad \text{for every } \mathbf{x}, \mathbf{y} \in R^n, \qquad (5.21)$$

where $L(f) > 0$ is called the Lipshitz constant of ∇f.

- $g : R^n \to R$ is a continuous convex function which is possibly *nonsmooth*.

For example, $f(\mathbf{x})$ can be a loss function and $g(\mathbf{x})$ a regularizer, such as in the case of the Lasso problem where $f(\mathbf{x}) = ||\mathbf{y} - \mathbf{A}\mathbf{x}||_2^2$ and $g(\mathbf{x}) = ||\mathbf{x}||_1$. Clearly, when $g(\mathbf{x}) = 0$, the problem in eq. 5.20 is reduced to the standard unconstrained smooth convex minimization problem.

Proximal methods are iterative algorithms that, starting with some initial point $\mathbf{x}_0$, compute a sequence of updates $\mathbf{x}_k$ that converges to the actual solution of (P). Given the current $\mathbf{x}_k$ obtained at iteration k, the next iterate $\mathbf{x}_{k+1}$ is found by minimizing the following quadratic approximation of the objective function

$$\min_{\mathbf{x} \in R^n} f(\mathbf{x}_k) + \nabla f(\mathbf{x}_k)^T(\mathbf{x} - \mathbf{x}_k) + \frac{L}{2}||\mathbf{x} - \mathbf{x}_k||_2^2 + g(\mathbf{x}),$$

where the first two terms constitute the linearization of f around the current point $\mathbf{x}_k$, and the third (quadratic) term, called the *proximal term*, helps to keep the next update $\mathbf{x}_{k+1}$ near the current iterate $\mathbf{x}_k$, where f is close to its linear approximations; $L > 0$ is a constant, which should be an upper bound on the Lipshitz constant $L(f)$, and is usually computed in practice via line-search. Using simple algebra, and removing constant terms independent of $\mathbf{x}$, we get the following *proximal problem*:

$$(PP): \quad \min_{\mathbf{x} \in R^n} \frac{1}{2}||\mathbf{x} - (\mathbf{x}_k - \frac{1}{L}\nabla f(\mathbf{x}_k))||_2^2 + \frac{1}{L}g(\mathbf{x}). \qquad (5.22)$$

We use the following notation:

$$G_f(\mathbf{x}) = \mathbf{x} - \frac{1}{L}\nabla f(\mathbf{x})) \qquad (5.23)$$

which denotes the standard *gradient update* step, as used in well-known gradient descent methods, while

$$Prox_{\mu g}(\mathbf{z}) = \arg\min_{\mathbf{x} \in R^n} \frac{1}{2}||\mathbf{x} - \mathbf{z}||_2^2 + \mu g(\mathbf{x}) \qquad (5.24)$$

Proximal Algorithm

1. **Initialize:**
   choose $\mathbf{x}_0 \in R^n$
   $L \leftarrow L(f)$, a Lipshitz constant of ∇f
   $k \leftarrow 1$

2. **Iteration step** k:
   $G_f(\mathbf{x}_{k-1}) \leftarrow \mathbf{x}_{k-1} - \frac{1}{L}\nabla f(\mathbf{x}_{k-1})$
   $\mathbf{x}_k \leftarrow Prox_{\mu g}(G_f(\mathbf{x}_{k-1}))$
   $k \leftarrow k+1$.

3. **If** a convergence criterion is satisfied, **then** exit, **otherwise** go to step 2.

FIGURE 5.10: Proximal algorithm.

denotes the *proximal operator* (here $\mu = \frac{1}{L}$), that is dating back to (Moreau, 1962), and to early proximal algorithms proposed by (Martinet, 1970) and (Lions and Mercier, 1979). Then the proximal method can be written concisely as a sequence of iterative updates, starting from $k = 0$ and an initial point $\mathbf{x}_0 \in R^n$:

$$\mathbf{x}_{k+1} \leftarrow Prox_{\mu g}(G_f(\mathbf{x}_k)). \tag{5.25}$$

Note that in the absence of the (nonsmooth) function $g(\mathbf{x})$, the proximal step reduces to the standard *gradient update* step

$$\mathbf{x}_{k+1} \leftarrow G_f(\mathbf{x}_k) = \mathbf{x}_k - \frac{1}{L}\nabla f(\mathbf{x}_k),$$

so that the proximal method reduces to the well-known *gradient descent* algorithm. Another standard technique that the proximal approach includes as a particular case is *projected gradient*: namely, if, for some set $S \subset R^n$, we assign $g(\mathbf{x}) = 0$ when $\mathbf{x} \in S$, and $g(\mathbf{x}) = +\infty$ when $\mathbf{x} \notin S$ (i.e., $g(\mathbf{x})$ is an indicator function for S), then solving the proximal problem becomes equivalent to computing a gradient update and projecting it on S, i.e. performing the projected gradient step:

$$\mathbf{x}_{k+1} \leftarrow Proj_S(G_f(\mathbf{x}_k)),$$

where $Proj_S$ is the projection operator on set S. An algorithmic scheme for the proximal methods is given in Figure 5.10.

5.3.3.2 Accelerated Methods

An attractive property of proximal algorithms is their simplicity. However, they may be slow to converge. The convergence of proximal methods is well-studied in

---

Accelerated Proximal Algorithm FISTA

1. **Initialize:**
   choose $\mathbf{y}_1 = \mathbf{x}_0 \in R^n$, $t_1 = 1$.
   $L \leftarrow L(f)$, a Lipshitz constant of ∇f
   $k \leftarrow 1$

2. *Iteration step* k: $G_f(\mathbf{x}_{k-1}) \leftarrow \mathbf{x}_{k-1} - \frac{1}{L}\nabla f(\mathbf{x}_{k-1})$
   $\mathbf{x}_k \leftarrow Prox_{\mu g}(G_f(\mathbf{y}_{k-1}))$
   $t_{k+1} = \frac{1+\sqrt{1+4t_k^2}}{2}$
   $\mathbf{y}_{k+1} = \mathbf{x}_k + (\frac{t_k-1}{t_{k+1}})(\mathbf{x}_k - \mathbf{x}_{k-1})$
   $k \leftarrow k+1$

3. **If** a convergence criterion is satisfied, **then** exit, **otherwise** go to step 2.

FIGURE 5.11: Accelerated proximal algorithm FISTA.

the literature – see, for example, (Figueiredo and Nowak, 2003; Daubechies et al., 2004; Combettes and Wajs, 2005; Beck and Teboulle, 2009) – and is known to be $O(\frac{1}{k})$, where k is the number of iterations, i.e.

$$F(\mathbf{x}_k) - F(\hat{\mathbf{x}}) \simeq O(\frac{1}{k}),$$

where $F(\mathbf{x}) = f(\mathbf{x}) + g(\mathbf{x})$, $\hat{\mathbf{x}}$ is the optimal solution of the problem (P) in eq. 5.20, and $\mathbf{x}_k$ is the k-th iterate of the proximal method.

Recently, several accelerated methods were proposed to improve the slow convergence of the basic proximal methods. The most prominent is FISTA, or Fast Iterative Shrinkage-Thresholding Algorithm of (Beck and Teboulle, 2009)[4]. It is similar to the basic proximal algorithm, but instead of the update step in eq. 5.26 it performs the following update:

$$\mathbf{x}_{k+1} \leftarrow Prox_{\mu g}(G_f(\mathbf{y}_k)),$$

where $\mathbf{y}_k$ is a specific linear combination of the two previous points $\mathbf{x}_{k-2}$ and $\mathbf{x}_{k-1}$, rather than simply the previous point $\mathbf{x}_{k-1}$. An algorithmic scheme for the FISTA method is given in Figure 5.11. A similar approach was proposed by Nesterov in 1983 (Nesterov, 1983) for minimization of smooth convex functions, and proven to be an *optimal* first-order, i.e. gradient, method in the complexity analysis sense as presented in (Nemirovsky and Yudin, 1983). FISTA extends Nesterov's algorithm to the case of nonsmooth objectives in eq. 5.20, improving the convergence rate from $O(\frac{1}{k})$ to $O(\frac{1}{k^2})$.

---

[4]Note that this algorithm extends the proximal methods in the general setting in eq. 5.20, not only in the specific case of the Lasso-solving ISTA method described in the next section, as its name may suggest.

5.3.3.3 Examples of Proximal Operators

An important step in determining the efficiency of the proximal approach is the computation of the proximal operator $Prox_{\mu g}$ in eq. 5.25. It turns out that the proximal operator can be computed efficiently, in closed form, for many popular types of the regularization function $g(\mathbf{x})$ involving l_q-norms. Herein, we briefly summarize these results, without their derivation; for more details, see (Combettes and Wajs, 2005) and (Bach et al., 2012). In the rest of this section we use the notation $\mu = \lambda/L$.

l_1-norm penalty (LASSO)

When the LASSO penalty $g(\mathbf{x}) = \lambda||\mathbf{x}||_1$ is used in eq. 5.20, the proximal problem (PP) in eq. 5.22 becomes

$$\min_{\mathbf{x} \in R^n} \frac{1}{2}||\mathbf{x} - G_f(\mathbf{x}_k)||_2^2 + \mu||\mathbf{x}||_1,$$

where $\mu = \lambda/L$. Note that the above problem is equivalent to eq. 5.4 in the previous section on univariate thresholding, where $\hat{\mathbf{x}} = G_f(\mathbf{x}_k)$. Namely, the problem decomposes into a set of univariate Lasso problems which are solved independently by the soft thresholding operator. Thus, the proximal operator in the case of an l_1-norm penalty is just the soft thresholding operator:

$$[Prox_{l_1}(\mathbf{x})]_i = (1 - \frac{\mu}{|x_i|})_+ x_i = sign(x_i)(|x_i| - \mu)_+.$$

Thus, for the l_1-norm regularizer $g(\mathbf{x})$, we recovered the popular Iterative Shrinkage-Thresholding Algorithm (ISTA), developed and analyzed independently by multiple researchers in different areas. One of the earlier versions of ISTA was presented in (Nowak and Figueiredo, 2001) and (Figueiredo and Nowak, 2003) as an expectation-maximization (EM) approach, and later in (Figueiredo and Nowak, 2005) in a majorization-minimization framework. In (Daubechies et al., 2004), the ISTA approach convergence results were derived. Iterative shrinkage-thresholding algorithms were also independently proposed around the same time by several other researchers (Elad, 2006; Elad et al., 2006; Starck et al., 2003a,b). In (Combettes and Wajs, 2005), a connection was made with a general class of forward-backward splitting algorithms, as discussed earlier.

l_2^2 penalty

The ridge-regression penalty $g(\mathbf{x}) = \lambda||\mathbf{x}||_2^2$ gives rise to the scaling operator:

$$Prox_{l_2^2}(\mathbf{x}) = \frac{1}{1+\mu}\mathbf{x}.$$

We will now mention two more examples of proximal operators, for the penalties we have not yet formally introduced: the Elastic Net and the group LASSO. However, both of those penalties will be discussed in significant detail very soon–in the next chapter.

$l_1 + l_2^2$ Elastic–Net penalty

When the Elastic–Net (Zou and Hastie, 2005) penalty is used, i.e. $g(\mathbf{x}) = \lambda(||\mathbf{x}||_1 + \alpha||\mathbf{x}||_2^2)$, where $\alpha > 0$, the proximal operator is obtained in a closed form as:

$$Prox_{l_1 + l_2^2}(\mathbf{x}) = \frac{1}{1 + 2\mu\alpha} Prox_{\mu||\cdot||_1}(\mathbf{x}).$$

l_1/l_2 group–LASSO penalty

Given a partition of the $\mathbf{x}$'s coordinates into J groups, the group–LASSO penalty (Yuan and Lin, 2006) corresponds to $g(\mathbf{x}) = \lambda \sum_{j=1}^{J} ||\mathbf{x}_j||_2$, where $\mathbf{x}_j$ is a projection of $\mathbf{x}$ on the j-th group. The proximal operator for this penalty is obtained in a closed form as follows:

$$[Prox_{l_1/l_2}(\mathbf{x})]_j = (1 - \frac{\mu}{||\mathbf{x}_j||_2})_+ \mathbf{x}_j, \quad j \in \{1, ...J\}.$$

In other words, this is the l_1-norm proximal operator applied at the group level, i.e. applied to each projection $\mathbf{x}_j$ on the j-th group, instead of each single coordinate.

For many more examples of proximal operators, including fused LASSO penalty (total variation) $\sum_{i=1}^{n-1} |x_{i+1} - x_i|$, where x_i is the i-th coordinate of $\mathbf{x}$, combined $l_1 + l_1/l_q$ norms, hierarchical l_1/l_q norms, overlapping l_1/l_∞ norms, as well as trace norms, see (Bach et al., 2012).

5.4 Summary and Bibliographical Notes

In this chapter, we discussed some popular algorithms for sparse signal recovery, such as approximate greedy search methods for the l_0-norm minimization, as well as several exact optimization techniques for the convex l_1-norm relaxation (LASSO problem), such as Least Angle Regression (LARS), coordinate descent, and proximal methods. An interesting feature of LARS is that it is a path-building algorithm, i.e. it produces the full set of LASSO solutions as the regularization parameter λ

decreases from infinity (empty solution) to zero (no sparsity). LARS is closely related to the homotopy method of (Osborne et al., 2000a). Note that path-building approaches were also developed for extensions of the LASSO problem, such as the Elastic Net (Zou and Hastie, 2005) and Generalized Linear Models (*glmpath* of (Park and Hastie, 2007)), discussed later in this book; also, general sufficient conditions for the piece-wise linearity of the path are discussed in (Rosset and Zhu, 2007). Similarly, the (block) coordinate descent and proximal methods are commonly used for solving sparse recovery problems that generalize LASSO to other loss functions and regularizers, discussed later in this book. See, for example, (Bach et al., 2012) and references therein for more details on these methods and for comprehensive empirical evaluation of those techniques over a variety of problems.

Clearly, it was not possible to fully cover the huge space of algorithms for sparse recovery that were developed recently, and continue being developed as we write this book. As noted before, the reader is referred to the Compressed Sensing Repository at http://dsp.rice.edu/cs, for a comprehensive set of references and links to available software packages implementing different approaches to sparse recovery.

Chapter 6

Beyond LASSO: Structured Sparsity

As it was already discussed, the LASSO method has two main advantages over the standard linear regression, namely, the l_1-norm regularization (1) helps to avoid overfitting the model on high-dimensional but small-sample data, and (2) facilitates embedded variable selection, i.e. finding a relatively small subset of relevant predictors. However, LASSO has also certain limitations that motivated development of several more advanced sparse methods in the recent years. In this chapter, we discuss some of these methods such as the Elastic Net, fused LASSO, group LASSO, and closely related to it, simultaneous LASSO (i.e., multi-task learning), as well as some practical applications of those methods. These methods focus on proper handling of correlated variables and various types of additional structure in practical applications and thus will be called *structured sparsity* methods[1].

In the following, we will assume that the response variable $\mathbf{y}$ is centered to have zero mean and all predictors are standardized to have zero mean and unit length (l_2-norm):

$$\sum_{i=1}^{m} y_i = 0, \ \sum_{i=1}^{m} a_{ij} = 0 \ \ and \ \ \sum_{i=1}^{m} a_{ij}^2 = 1, \ 1 \le j \le n.$$

---

[1] In the literature, structured sparsity typically refers to group LASSO and similar methods, but here we use this terminology in a broader sense, including all methods that take into account interactions among the predictors.

6.1   The Elastic Net

We will first discuss the *Elastic Net* regression method of (Zou and Hastie, 2005). This method was primarily motivated by applications such as computational biology, where the predictor variables tend to be correlated with each other, and often form groups, or clusters, of similarly relevant (or irrelevant) variables with respect to predicting the target. As argued in (Zou and Hastie, 2005), the original LASSO may not be ideal for such applications due to the following issues:

- When the number of variables n exceeds the number of observations m, and the LASSO solution is unique (which happens with probability one when the entries of $\mathbf{A}$ are drawn from a continuous probability distribution (Tibshirani, 2013)), it contains at most m nonzero coefficients (Osborne et al., 2000b). This limitation can be undesirable when the main goal is identification of important predictive variables, since the number of such variables may exceed m.

- If a group of correlated predictors is highly relevant to the target variable, it is desirable to include all such predictors in a sparse model with *similar coefficients*, and particularly, equal predictors must have equal coefficients in the model. However, this is not necessarily the case with the LASSO solutions, i.e. they lack the desired *grouping property*. As discussed in (Zou and Hastie, 2005), it was observed empirically that LASSO tends to select one (arbitrary) variable from a group of highly correlated ones.

- Also, as it was observed in the original LASSO paper (Tibshirani, 1996), empirical predictive performance of LASSO is dominated by ridge regression (i.e. l_2-norm-regularized linear regression) when the predictors are highly correlated, while the situation is reversed when there is a relatively small number of more independent variables. Thus combining both l_1- and l_2-norms may be necessary to achieve "the best of both worlds".

In particular, let us consider the grouping property. Indeed, when two predictors are exactly the same, one would expect a linear model to assign equal coefficients to both variables. As discussed in (Zou and Hastie, 2005), this is achieved by penalized linear regression

$$\min_{\mathbf{x}} \frac{1}{2} ||\mathbf{y} - \mathbf{A}\mathbf{x}||_2^2 + \lambda R(\mathbf{x}) \tag{6.1}$$

when the penalty $R(\mathbf{x})$ is *strictly convex* (see chapter 2 for the definition of strict convexity); note that it is also always assumed that the penalty $R(\mathbf{x}) > 0$ for $\mathbf{x} \neq 0$. However, since the l_1-norm is convex, but not strictly convex, this property is not satisfied by the LASSO, as summarized in the following:

Lemma 6.1. *(Zou and Hastie, 2005) Let $\mathbf{a}_i = \mathbf{a}_j$, where $\mathbf{a}_i$ and $\mathbf{a}_j$ are the i-th and j-th columns of the design matrix $\mathbf{A}$, let $R(\mathbf{x}) > 0$ for $\mathbf{x} \neq 0$, and let $\hat{\mathbf{x}}$ be a solution of 6.1.*

- If $R(\mathbf{x})$ is strictly convex, then $x_i = x_j$ for any $\lambda > 0$.

- If $R(\mathbf{x}) = ||\mathbf{x}||_1$, then $x_i x_j \geq 0$, and there are infinitely many solutions $\mathbf{x}'$ where $x_i' = \alpha(\hat{x}_i + \hat{x}_j)$ and $x_j' = (1 - \alpha)(\hat{x}_i + \hat{x}_j)$, for any $0 \leq \alpha \leq 1$, while the rest of the coefficients remain the same, i.e. $x_k' = \hat{x}_k$ for all $k \neq i$, $k \neq i$.

In order to avoid the above issue, (Zou and Hastie, 2005) propose the method called the *Elastic Net (EN)*, which augments LASSO's regularization with an additional squared l_2-norm term that makes the regularization function strictly convex. More precisely, the following optimization problem

$$\min_{\mathbf{x}} ||\mathbf{y} - \mathbf{A}\mathbf{x}||_2^2 + \lambda_1||\mathbf{x}||_1 + \lambda_2||\mathbf{x}||_2^2 \qquad (6.2)$$

is called the *naive* Elastic Net problem, and the (corrected) Elastic Net solutions is simply a re-scaled version of the solution to the above naive Elastic Net, as originally proposed in (Zou and Hastie, 2005); we will discuss the re-scaling issue in more detail at the end of this section. It is easy to see from eq. 6.2 that the naive Elastic Net becomes equivalent to the LASSO when $\lambda_2 = 0$ and $\lambda_1 > 0$, while for $\lambda_1 = 0$ and $\lambda_2 > 0$ it becomes equivalent to the ridge regression (Hoerl and Kennard, 1988), i.e. the l_2-norm regularized linear regression. Clearly, when both λ_1 and λ_2 are zero, the naive Elastic Net problem simply reduces to the ordinary least-squares linear regression, or OLS.

The geometry of the Elastic Net penalty in the two-dimensional case is shown in Figure 6.1; it can be also viewed as a convex combination of the LASSO and ridge penalties:

$$\alpha||\mathbf{x}||_2^2 + (1 - \alpha)||\mathbf{x}||_1, \quad \text{where } \alpha = \frac{\lambda_2}{\lambda_1 + \lambda_2}.$$

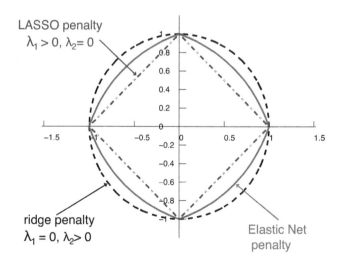

FIGURE 6.1: Contour plots for the LASSO, ridge, and Elastic Net penalties at the same function value of 1.

Note that the Elastic Net penalty still has sharp corners (singularities) that enforce sparse solutions, similarly to the l_1-norm penalty of the LASSO. However, unlike the LASSO penalty, the Elastic Net penalty is strictly convex and thus guarantees that equal predictors will be assigned equal coefficients. The Elastic Net can be easily transformed into an equivalent LASSO problem over a modified ("augmented") data matrix, and solved by standard methods for solving LASSO, as shown in the following.

Lemma 6.2. *(Zou and Hastie, 2005) Let* $(\mathbf{y}^*, \mathbf{A}^*)$ *be an "augmented" dataset defined as follows:*

$$\mathbf{A}^*_{(m+n)\times n} = \frac{1}{\sqrt{1+\lambda_2}} \begin{pmatrix} \mathbf{A} \\ \sqrt{\lambda_2}\mathbf{I} \end{pmatrix}, \quad \mathbf{y}^*_{(m+n)} = \begin{pmatrix} \mathbf{y} \\ 0 \end{pmatrix},$$

and let $\gamma = \lambda_1/\sqrt{1+\lambda_2}$ *and* $\mathbf{x}^* = \sqrt{1+\lambda_2}\mathbf{x}$. *Then the naive Elastic Net problem can be written as*

$$\min_{\mathbf{x}^*} ||\mathbf{y}^* - \mathbf{A}^*\mathbf{x}^*||_2^2 + \gamma||\mathbf{x}^*||_1. \tag{6.3}$$

If $\hat{\mathbf{x}}^*$ *is the solution of the above problem 6.3, then*

$$\hat{\mathbf{x}} = \frac{1}{\sqrt{1+\lambda_2}}\hat{\mathbf{x}}^*$$

is the solution to the naive Elastic Net problem in eq. 6.2.

In other words, solving the naive Elastic Net problem is equivalent to solving the LASSO problem in eq. 6.3 with the regularization weight $\gamma = \lambda_1/\sqrt{1+\lambda_2}$.

Since the number of samples in the problem stated in eq. 6.3, over the augmented dataset, is $m^* = m + n$ and since $\mathbf{A}^*$ has full column-rank, the Elastic Net solution can include up to n predictors, i.e. all of them, thus eliminating one of the limitations of the basic LASSO approach. As we already mentioned above, the (naive) Elastic Net regularizer is strictly convex, implying that equal variables are assigned equal coefficients, and thus eliminating another drawback of LASSO. Finally, as the following result shows, the (naive) Elastic Net penalty enforces the desirable *grouping* of highly correlated variables.

Theorem 6.3. *(Zou and Hastie, 2005) Let* $\hat{\mathbf{x}}(\lambda_1, \lambda_2)$ *be the solution of the naive Elastic Net problem in eq. 6.2 for given* λ_1 *and* λ_2; *as usual, it is assumed that* $\mathbf{y}$ *is centered and columns of* $\mathbf{A}$ *are standardized. Let* $\hat{\mathbf{x}}_i(\lambda_1, \lambda_2) > 0$, *and let the (normalized) absolute difference between the* i-*th and* j-*th coefficients be defined as*

$$d_{\lambda_1,\lambda_2}(i,j) = \frac{1}{||\mathbf{y}||_1}|\hat{x}_i(\lambda_1, \lambda_2) - \hat{x}_i(\lambda_1, \lambda_2)|.$$

Then

$$d_{\lambda_1,\lambda_2}(i,j) \leq \frac{1}{\lambda_2}\sqrt{2(1-\rho)},$$

where $\rho = \mathbf{a}_i^T \mathbf{a}_j$ *is the sample correlation between the* i-*th and* j-*th predictors.*

In other words, higher correlation between two predictors implies higher similarity, i.e. smaller difference, between their coefficients, provided that the coefficients are of the same sign. This property is often referred to as *grouping* of similar predictors. Since the λ_2 parameter facilitates such grouping effect, we will also refer to it as the *grouping* parameter, while the λ_1 weight on the l_1-norm is often referred to as the *sparsity* parameter. Note that $d_{\lambda_1,\lambda_2}(i,j)$ is inversely proportional to λ_2, and thus coefficients of the same-sign variables in the naive Elastic Net solution become more similar with increasing λ_2. However, in the (corrected) Elastic Net solution, as we discuss later, they will be rescaled by $(1 + \lambda_2)$, and thus the difference between the coefficients will be only controlled by ρ, the correlation between the corresponding predictors.

Since the (naive) Elastic Net problem is equivalent to the LASSO problem on an augmented dataset, it can be solved by any method for solving LASSO. In (Zou and Hastie, 2005), the procedure called LARS-EN is proposed. This is a relatively simple modification (that exploits sparse structure of the augmented data matrix) of the popular algorithm for solving LASSO, Least Angle Regression using a Stagewise procedure (LARS) (Efron et al., 2004). LARS-EN has two input parameters: the *grouping* parameter λ_2 and the *sparsity* parameter k that specifies the maximum number of *active* predictors, i.e. the predictors having nonzero coefficients in $\hat{\mathbf{x}}$ (also called the *active set*). It can be shown (Efron et al., 2004) that each value of k corresponds to a unique value of λ_1 in eq. 6.2, with larger λ_1 (i.e., larger weight on l_1-norm penalty) enforcing sparser solutions and thus corresponding to a *smaller* number of nonzero coefficients k. LARS-EN produces a collection of solutions, i.e. the regularization path, for all values of k varying from 1 to a specified maximum number of nonzeros. As such, the sparsity parameter is also referred to as the *early stopping* parameter, since it serves as a stopping criterion for the LARS-EN incremental procedure; for more details on the LARS procedure, see chapter 5. Note that LARS-EN, like the original LARS, is highly efficient, as it finds the entire regularization path at the cost of a single OLS fit. Remember that knowing the regularization path facilitates choosing the most-predictive solution $\mathbf{x}$ and its corresponding parameter λ_1 using cross-validation.

Finally, the Elastic Net estimate of the linear regression coefficients is obtained by rescaling the naive Elastic Net solution. As argued in (Zou and Hastie, 2005), such rescaling is needed to compensate for double-shrinkage of the coefficients due to combination of both LASSO and ridge penalties in the naive Elastic Net. While the naive Elastic Net achieves desired properties as a variable-selection approach, i.e., it overcomes LASSO's limitation on the number of nonzeroes and promotes grouping of correlated variables, its predictive performance can be inferior as compared to the LASSO or ridge regression; (Zou and Hastie, 2005) argue that the reason for this could be the double-shrinkage, or over-penalizing, of the coefficients due to combination of penalties in the Elastic Net, as compared to single penalties of the LASSO and ridge. As a solution to this double-shrinkage problem, (Zou and Hastie, 2005) propose to use the *(corrected) Elastic Net* estimate defined as rescaled version of the naive Elastic Net solution, $\hat{\mathbf{x}}(EN) = \sqrt{1 + \lambda_2}\hat{\mathbf{x}}^*$. By Lemma 6.2,

$\hat{\mathbf{x}}(naive\ EN) = \frac{1}{\sqrt{1+\lambda_2}}\hat{\mathbf{x}}^*$, and thus

$$\hat{\mathbf{x}}(EN) = (1 + \lambda_2)\hat{\mathbf{x}}(naive\ EN),$$

i.e. *the Elastic Net solution is a rescaled version of the naive Elastic Net solution.*

In summary, this section described the Elastic Net (EN), a regularized linear regression method that combines the l_1- and l_2-norm regularizers and

- is equivalent to LASSO when $\lambda_2 = 0$ and to ridge regression when $\lambda_1 = 0$; also, as shown in eq. 16 in Zou and Hastie (2005), when $\lambda_2 \to \infty$ the Elastic Net becomes equivalent to *univariate soft thresholding*, i.e. to correlation-based variable selection method with a particular threshold value;

- EN overcomes LASSO's limitation on the number of nonzero predictors, i.e. can potentially choose up to n nonzeros;

- EN enforces solution sparsity while also encouraging grouping effect, i.e. similar coefficients among correlated predictors; equal predictors are guaranteed to be assigned equal coefficients.

6.1.1 The Elastic Net in Practice: Neuroimaging Applications

We will now discuss practical applications of the Elastic Net, focusing on some high-dimensional prediction problems arising in neuroimaging, and particularly, in the analysis of functional magnetic resonance images (fMRIs). As discussed earlier in chapter 1, one of the common questions in fMRI analysis is discovering brain areas (i.e. subsets of voxels) that are relevant to a given stimulus, task, or mental state. However, the traditional mass-univariate GLM approach based on individual voxel activations (correlations between each voxel and the stimulus, task, or mental state) often misses potentially informative voxel interactions (Haxby et al., 2001; Rish et al., 2012b). Thus, multivariate predictive modeling, and sparse modeling in particular, become increasingly popular analysis tools in neuroimaging.

For example, the Elastic Net was shown to produce surprisingly accurate predictive models of pain perception (Rish et al., 2010). Brain imaging analysis of pain is a rapidly growing area of neuroscience, motivated both by a scientific goal of improving our understanding of pain mechanisms in the human brain and by practical medical applications (Baliki et al., 2009, 2008). While most of the literature on pain is still focused on univariate analysis, a more recent work explores the advantages of sparse modeling for predicting a subject's pain perception from his or her fMRI data (Rish et al., 2010), as well as better characterization of brain areas relevant to pain processing (Rish et al., 2012b). The results discussed in (Rish et al., 2010, 2012b; Cecchi et al., 2012) were obtained on an fMRI dataset originally presented in (Baliki et al., 2009). A group of 14 healthy subjects participated in this study. The subjects in the scanner were asked to rate their pain level using a finger-span device in response to a series of painful thermal stimuli applied to their back. (An fMRI-compatible device was used to deliver fast-ramping painful thermal stimuli via a contact probe.)

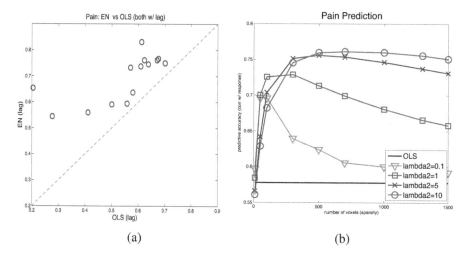

FIGURE 6.2: (a) Predictive accuracy of the Elastic Net vs. OLS regression for the task of predicting thermal pain perception from fMRI data; (b) effects of the sparsity and grouping parameters on the predictive performance of the Elastic Net.

The task addressed in (Rish et al., 2010, 2012b) was to learn a regression model that could predict a subject's pain rating based on his/her fMRI data. The individual time slices (at which "brain snapshots" were taken) correspond to samples (rows in **A**), and the voxels correspond to predictors (columns in **A**). The target variable **y** here is the pain perception level as rated by a subject, at a particular point in time.

Figure 6.2, reproduced here from (Rish et al., 2010), shows the predictive performance of the Elastic Net (EN) regression measured by correlation between the predicted and actual pain rating by a subject. We can see that, first of all, EN always outperforms unregularized linear regression (OLS), often by a large margin, as shown in panel (a); EN often achieves up to 0.7-0.8 (and never below 0.5) correlation between the predicted and actual pain ratings. The results here are shown for EN with a fixed sparsity (1000 active predictors) and grouping ($\lambda_2 = 20$). On the other hand, panel (b) compares EN performance at different levels of sparsity and grouping, for one of the subjects from the group (the results were typical for the whole group). We can see that the EN's prediction improves with growing λ_2 (recall that smaller values such as $\lambda_2 = 0.1$ make EN more similar to LASSO). Note that the optimal (prediction-wise) number of active variables selected by EN at each fixed λ_2 increases with increasing λ_2. Apparently, at higher values of the grouping parameter, more (correlated) variables are pulled into the model and jointly provide a more accurate prediction of the target variable, pain perception.

Another example of a successful use of the Elastic Net in fMRI analysis is predicting mental states of subjects playing a videogame (Carroll et al., 2009), based on the dataset provided by the 2007 Pittsburgh Brain Activity Interpretation Competition Pittsburgh EBC Group (2007). There were more than 20 different response variables in this dataset, including some "objective" and "subjective" variables. The objective response variables, such as picking up certain objects or hearing instructions during the game, were measured simultaneously with the functional MRI data, while a few subjective response variables (e.g., feeling anxious) were estimated off-line based on videogame recording: namely, a subject was shown a recording of the videogame just played, and asked to rate particular aspects of his or her emotional state.

One of the key observation in (Carroll et al., 2009) is that tuning the grouping parameter λ_2 allows to achieve such useful characteristics of a model as *better interpretability* and *better stability*. In neuroscience, and other biological applications, such qualities are playing a particularly important role, since the ultimate goal of statistical data analysis in such applications is to shed light on underlying natural phenomena and guide scientific discovery. In the experiments summarized below, for each value of λ_2, the optimal sparsity level (i.e. the desired number of nonzero coefficients in the LARS-EN procedure) was selected by cross-validation. Figure 6.3

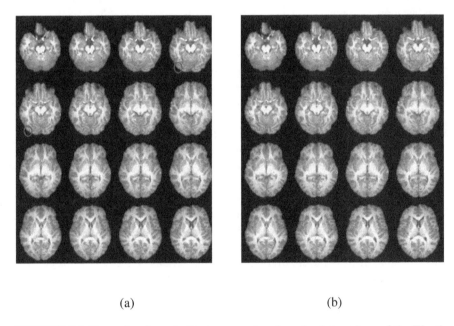

(a)   (b)

FIGURE 6.3 (See color insert): Brain maps showing absolute values of the Elastic Net solution (i.e. coefficients x_i of the linear model) for the Instruction target variable in PBAIC dataset, for subject 1 (radiological view). The number of nonzeros (active variables) is fixed to 1000. The two panels show the EN solutions (maps) for (a) $\lambda_2 = 0.1$ and (b) $\lambda_2 = 2$. The clusters of nonzero voxels are bigger for bigger λ_2, and include many, but not all, of the $\lambda_2 = 0.1$ clusters. Note that the highlighted (red circle) cluster in (a) is identified by EN with $\lambda_2 = 0.1$, but not in the $\lambda_2 = 2.0$ model.

visualizes EN models trained on one of the subject's fMRI data for predicting one of the target variables, called "Instructions", which indicates the beginning of an auditory playback of instructions, repeated regularly during the videogame. The absolute values of the regression coefficients corresponding to brain voxels are shown on brain maps (each map corresponds to a different horizontal slice of a 3D MRI image), with nonzero voxels highlighted in color. Two levels of λ_2, low ($\lambda_2 = 0.1$) and high ($\lambda_2 = 2.0$), are compared in panels (a) and (b), respectively. We can see that the higher value of the grouping parameter produces somewhat larger, spatially contiguous clusters while the solutions for lower λ_2 (i.e., closer to LASSO) are sparser and more spotty. From the point of view of neuroscientific interpretation, spatially contiguous clusters of voxels related to a given task or stimulus (e.g., listening to instructions) make more sense that spotty maps, even though both models can be equally predictive (see Figure 6.4a). This is because the BOLD signals of neighboring voxels are typically highly correlated as they reflect the blood flow to that particular brain region, and thus the whole region, or cluster, should be shown as "active", or relevant, for a given task. As discussed earlier (Zou and Hastie, 2005), LASSO would not be an ideal variable-selection method here since it tends to choose one "representative" variable from a cluster of correlated variables, and ignore the rest. On the contrary, the Elastic Net is indeed demonstrating grouping property desirable for making sparse solutions more interpretable, and controlling the grouping parameter λ_2 helps to improve interpretability of the model.

Besides improving the interpretability of the model, tuning λ_2 can also improve its *stability* (also called *robustness*), as demonstrated in (Carroll et al., 2009). In order to measure stability, the following quantities were computed: V_{total}, the total number of unique voxels selected by EN over the two experimental runs for each subject, and V_{common}, the number of voxels co-occurring in the two models. The stability was then computed as the ratio V_{common}/V_{total}. It was hypothesized that greater inclusion of voxels from within correlated clusters would result in greater overlap in included voxels between two models generated on different datasets. The results presented in Figure 6.4 (reproduced here from (Carroll et al., 2009)) indeed confirm this hypothesis. Though the prediction quality remains essentially the same (or slightly improved) when increasing the λ_2 value from 0.1 to 2.0 (Figure 6.4a), the stability of a model improves considerably, for practically all target variables (responses or stimuli) and all subjects (Figure 6.4b). Moreover, as Figure 6.4c shows, increasing λ_2 is frequently associated with the inclusion of a greater number of voxels (recall that the number of selected voxels is found via cross-validation). These additional voxels are likely those relevant voxels highly correlated with other relevant voxels. Hence, by including more relevant yet correlated voxels, increasing λ_2 improves model stability without compromising prediction performance.

Finally, sparse regression can be a much better tool than the traditional univariate GLM approach for discovering task-relevant brain activations. Indeed, GLM would essentially rank the voxels by their individual correlations with the task, and present the top most-correlated ones that survive statistical significance test as "brain activation" relevant to the task. However, predictive accuracy of subsets of voxels can be a better proxy for estimating which brain areas are relevant, or informative, about the

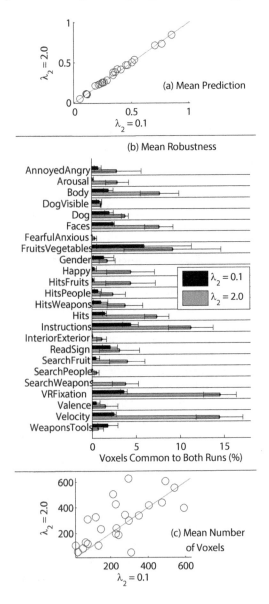

FIGURE 6.4: Even among equally predictive, cross-validated, models (a), increasing the λ_2 parameter increases model stability (b) while slightly increasing the number of included voxels (c). (a) Mean model prediction performance measured by correlation with test data. Prediction scores for $\lambda_2 = 2$ are plotted against their matching scores obtained with $\lambda_2 = 0.1$. (b) Stability of sparse models at different values of λ_2, for different target variables. (c) The mean number of voxels selected when a lower λ_2 value (0.1) is used are plotted against the matching mean number of voxels selected when a higher λ_2 value (2.0) is used. Mean values in (a) and (c) are computed over the 3 subjects and 2 experiments (runs), in (b) over the 3 subjects.

task. Moreover, it provides a tool for exploring multiple relevant sparse solutions. Indeed, given a brain map of task-relevant voxels either constituting a solution of the Elastic Net, or the top-ranked voxels found by GLM, what can be said about the remaining voxels, not shown on the map? Are they completely irrelevant? Or, vice versa, can such suboptimal voxels can be still highly relevant to a given task or mental state? As it turns out, the answer to the last question is positive for many tasks, as shown in (Rish et al., 2012b), which raises the question about the validity and limitations of standard brain-mapping approaches.

It is well-known that multiple near-optimal sparse solutions are possible in the presence of highly correlated predictors, and exploring the space of such solutions is an interesting open question. For example, a very simple procedure is used in (Rish et al., 2012b) to explore this space, by first finding the best EN solution with 1000 voxels, removing those voxels from the set of predictors, and repeating the procedure until there are no more voxels left. We will call the Elastic Net solutions obtained in such manner "restricted" since they are found on a restricted subset of voxels. Figure 6.5a plots the predictive accuracy of subsequent restricted solutions for the pain perception. Here the x-axis shows the total number of voxels used so far by the first k "restricted" solutions, with increments of 1000, since this is the size of each subsequent solution, found by the Elastic Net after removing from consideration all voxels from the previous solutions. Predictive accuracy (y axis) is measured by the correlation between the actual value (e.g., pain rating) and predicted value, computed on a test dataset separate from the training dataset (herein, the first 120 TRs were used for training, and the remaining 120 TRs for testing). It is quite surprising to see a very slow degradation of the predictive performance without any clear transition from highly predictive to irrelevant voxels, as it would be suggested by the "classical" brain map approach. In other words, for pain perception, there is *no clear separation between relevant and irrelevant areas*, suggesting that the task relevant information can be *widely distributed throughout the brain*, rather than localized in a small number of specific areas – the "holographic" effect, as it was referred to in (Rish et al., 2012b).

Note that the standard GLM method does not reveal such phenomenon, since, as shown in Figure 6.6 reproduced here from (Rish et al., 2012b), the individual voxel-task correlations always seem to decay exponentially (Figure 6.6b), and for many reasonably predictive (though not necessarily the top) sparse solutions, their voxel would not even pass the 0.1 correlation threshold (Figure 6.6b). For example, voxel correlations in the 10th and especially in the 25th solution are mostly below 0.3 and 0.1, respectively; however, the predictive power of those solutions (around 0.55) is still pretty good as compared to the best predictive accuracy obtained by the 1st solution (around 0.67). Thus, such results provide further support to the earlier observations by (Haxby et al., 2001) that *highly predictive models of mental states can be built from voxels with sub-maximal task correlations.* Moreover, these results suggest that *multivariate sparse models can provide a better tool than the standard mass-univariate neuroimaging methods for exploring the task-related information spread through the brain.*

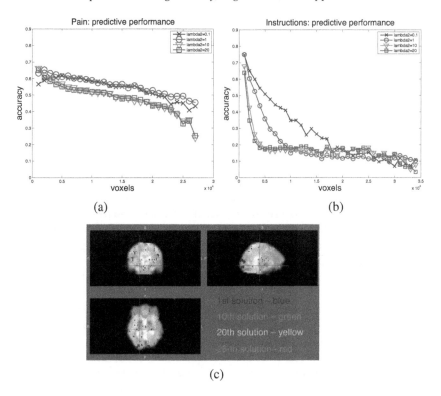

(a)   (b)

(c)

FIGURE 6.5 (See color insert): Predictive accuracy of the subsequent "restricted" Elastic Net solutions, for (a) pain perception and (b) "Instructions" task in PBAIC. Note very slow accuracy degradation in the case of pain prediction, even for solutions found after removing a significant amount of predictive voxels, which suggests that pain-related information is highly distributed in the brain (also, see the spatial visualization of some solutions in Figure (c)). The opposite behavior is observed in the case of the "Instruction" – a sharp decline in the accuracy after a few first "restricted" solutions are deleted, and very localized predictive solutions shown earlier in Figure 6.3.

   While several other tasks presented in (Rish et al., 2012b), besides the pain rating, showed a similar "holographic" effect, one task was different as it demonstrated fast (exponential) performance degradation, and clear separation of relevant vs. irrelevant areas; this was a relatively simple auditory task from PBAIC (Figure 6.5b). A possible hypothesis here can be that, while "simple" tasks are localized, more complex tasks/experiences (such as pain) tend to involve much more distributed brain areas.
   Also, note that the grouping property of the Elastic NET again proves useful here. Indeed, using high values of grouping parameter in the Elastic Net is essential to enforce inclusion of task-relevant clusters of voxels *together* (i.e. in the same solution), and prevent subsequent solutions from drawing voxels from the relevant areas already "used" by previous solutions, as the LASSO tends to do. This way, if

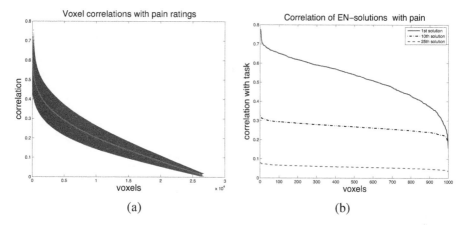

FIGURE 6.6: (a) Univariate correlations between each voxel and the pain ratings, sorted in decreasing order and averaged over 14 subjects. The line corresponds to the average, while the band around it shows the error bars (one standard deviation). Note that degradation of univariate voxel correlations is quite rapid, unlike the predictive accuracy over voxel subsets shown in Figure 1. (b) Univariate correlations with the pain rating for a single subject (subject 6th), and for three separate sparse solutions: the 1st, the 10th, and the 25th "restricted" solutions found by the Elastic net with $\lambda_2 = 20$.

there is indeed a clear separation between relevant versus irrelevant groups of voxels, the Elastic Net will detect it, as it happens with the Instructions task in the PBAIC dataset, while the LASSO (or the close-to-the-LASSO Elastic Net with small grouping parameter) will not.

6.2 Fused LASSO

As we just discussed, the Elastic Net generalizes the LASSO and builds regression models that combine two desirable properties: sparsity (due to the l_1-norm regularizer) and grouping, or coefficient similarity/smoothing, of highly-correlated predictors (due to the l_2-norm regularizer). Such smoothing of the model coefficients, however, does not use any other information about the problem structure that can be present in a specific application.

For instance, predictors may follow some natural *ordering*, with the coefficients changing smoothly along the ordering. Examples include temporal, spatial, or spatio-temporal orderings, arising in various signal-processing applications (e.g., imaging, audio and video processing). Other examples include protein mass-spectrography data and gene expression data, as discussed in (Tibshirani et al., 2005). In some applications, such as gene-expression analysis, the order may not be known in advance, but can be estimated from the data, e.g., via hierarchical clustering.

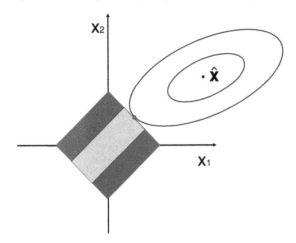

FIGURE 6.7: Geometry of the fused LASSO problem in two dimensions. The solution corresponds to the point where the contours of the quadratic loss function (the ellipses) touch the feasible region (the rectangle), i.e. the set of all points satisfying both the bounded l_1-norm constraint $||\mathbf{x}||_1 \leq t_1$ and the bounded-difference constraint $\sum_{i=1}^{n-1} |x_{i+1} - x_i| \leq t_2$.

In order to enforce smoothness along an ordering of predictors while building a sparse model, an approach called *fused LASSO* was proposed by (Tibshirani et al., 2005); besides the l_1-norm penalty on the model coefficients, it adds an l_1-norm penalty on the *differences* between the coefficients of successive predictors. The latter penalty encourages sparsity of such differences, i.e. "local constancy" of the coefficients along the given ordering:

$$\min_{\mathbf{x}} ||\mathbf{y} - \mathbf{Ax}||_2^2 \text{ subject to } ||\mathbf{x}||_1 \leq t_1 \text{ and } \sum_{i=1}^{n-1} |x_{i+1} - x_i| \leq t_2, \qquad (6.4)$$

or, equivalently, using Lagrangian multipliers:

$$\min_{\mathbf{x}} ||\mathbf{y} - \mathbf{Ax}||_2^2 + \lambda_1 ||\mathbf{x}||_1 + \lambda_2 \sum_{i=1}^{n-1} |x_{i+1} - x_i|. \qquad (6.5)$$

Figure 6.7 shows the geometry of fused LASSO in a two-dimensional case.

Fused LASSO has multiple applications, for example, in bioinformatics (Tibshirani and Wang, 2008; Friedman et al., 2007a). Also, fused LASSO is closely related to the *total variation (TV)* denoising in signal processing. For example, in case of one-dimensional signals, the only difference between the TV and fused Lasso problem formulations is that TV includes only the second penalty used by fused LASSO, i.e., the absolute difference, or variation, between the successive coordinates in **x**, and omits the l_1-norm penalty on **x**. The total variation problem has a long history in signal and image processing as a popular tool for image denoising (Rudin et al., 1992;

Blomgren and Chan, 1998; Chan and Shen, 2005). As it was shown in (Friedman et al., 2007a), the solution of the fused LASSO problem can be obtained by simple soft-thresholding of the TV solution, and thus both approaches have practically the same computational complexity, and an be solved by similar methods.

Moreover, besides its use in linear regression, the fused LASSO type of penalty was used for other loss functions – for example, to learn a sparse and locally-constant Gaussian Markov Random Field (MRF) (Honorio et al., 2009), with applications in fMRI analysis, where local constancy of model coefficients is implied by the spatial contiguity of the BOLD response signal, i.e. by the fact that brain activations usually involve not just a single voxel but the whole brain area around the voxel.

6.3 Group LASSO: l_1/l_2 Penalty

The fused LASSO and the Elastic Net impose additional constraints on the model parameters, such as similarity of highly-correlated coefficients or local constancy/similarity of subsequent coefficients along a given ordering. A closely related approach, called the *group LASSO*, also imposes an additional structural constraint besides the basic sparsity: it assumes there are groups of predictors that are naturally associated with each other, and thus must be included into (or excluded from) the model simultaneously. Knowing the groups in advance differentiates the group LASSO from the Elastic Net, where grouping of the predictors is a completely data-driven, correlation-based process. The group LASSO approach was introduced independently, for different applications and under different names, by several researchers – see, for example, (Bakin, 1999; Antoniadis and Fan, 2001; Malioutov et al., 2005; Lin and Zhang, 2006) – and also further generalized and analyzed in the work by (Yuan and Lin, 2006).

There are many applications with a natural group structure among the variables that can be used by group LASSO, such as, for example, a functional cluster of genes in DNA microarrays, or a set of voxels from the same brain areas in fMRI analysis. Another common example includes a group of binary indicator variables that encode categorical variables, or a group of parameters associated with the same predictor variable but multiple target variables in *multi-task learning*, as it will be discussed later.

Let us assume that the set $\mathbf{A} = \{A_1, ..., A_n\}$ of all predictor variables is partitioned into a set $G = \{G_1, ..., G_J\}$ of J (non-intersecting) groups, i.e.

$$\mathbf{X} = \bigcup_{j=1}^{J} G_j,$$

where $G_i \bigcap G_j = \emptyset$ for $i \neq j$. Let $\mathbf{x}_{G_j}$ denote a vector of linear coefficients that correspond to the group $g \in G$. The group LASSO problem is stated as follows

(Yuan and Lin, 2006):

$$\min_{\mathbf{x}} ||y - \mathbf{A}\mathbf{x}||_2^2 + \lambda \sum_{j=1}^{J} c_j ||\mathbf{x}_{G_j}||_2, \qquad (6.6)$$

where $\sum_{j=1}^{J} c_j ||\mathbf{x}_{G_j}||_2$ is called the *block* l_1/l_2 penalty. The weight c_j accounts for varying group sizes, and is usually set to $\sqrt{n_j}$, where n_j is the size of the group G_j. The block penalty uses the l_2-norm $||\mathbf{x}_{G_j}||_2$ within each group, and the l_1-norm across the groups, which is simply the sum of all $||\mathbf{x}_{G_j}||_2$, since l_2-norms are non-negative. Note that the group LASSO is equivalent to the standard LASSO when the groups consist of single variables, i.e., $G_j = \{A_j\}, 1 \leq j \leq n$, and thus $n_j = 1$ and $\sum_{j=1}^{J} c_j ||\mathbf{x}_{G_j}||_2 = \sum_{j=1}^{n} |x_j| = ||\mathbf{x}||_1$. In general, when the groups contain more than one variable, the group-level l_1-norm *encourages sparsity across the groups*, i.e. selection of a relatively small subset of groups from G, and setting to zeros *all* coefficients in the remaining groups. On the other hand, the l_2-norm over the coefficients in each group *discourages sparsity within each group*; namely, if a group is selected, then all variables in that group tend to have nonzero coefficients, i.e. are selected together. In summary, the group LASSO works as a standard LASSO at a group level, and the sparsity parameter λ controls how many groups are selected to have nonzero coefficients in the model.

Note that sometimes the weight factor $\sqrt{n_j}$ in eq. 6.6 is omitted, i.e. the group LASSO penalty is simply $\sum_{j \in J} ||\mathbf{x}_{G_j}||_2$. Both formulations are just particular cases of a more general group-enforcing penalty proposed by (Bakin, 1999):

$$\sum_{j=1}^{J} ||\mathbf{x}_j||_{K_j}, \qquad (6.7)$$

where $K_1, ..., K_J$ are positive definite matrices, and $||\mathbf{x}||_K = \sqrt{\mathbf{x}^T K \mathbf{x}}$, with $\mathbf{x}^T$ denotes the transpose of $\mathbf{x}$. The formulation in (Yuan and Lin, 2006) simply uses $K_j = n_j I_{n_j}$, where I_{n_j} is the identity matrix of size n_j; the no-weight formulation corresponds to $K_j = I_{n_j}$.

6.4   Simultaneous LASSO: l_1/l_∞ Penalty

Another approach to grouping variables was proposed by (Turlach et al., 2005), where the task of interest involved selecting a common subset of predictors when simultaneously predicting several target (response) variables. The task is called *simultaneous variable selection*, and is an extension of the LASSO to the case of multiple responses. Each group here consists of the linear regression parameters associated with a particular predictor across all regression tasks.

More specifically, we assume that there are k response variables, $Y_1,...,Y_k$, that we are going to predict from a set of n predictor variables $A_1,...,A_n$, and that only a relatively small subset of the *same* predictors is relevant to *all* responses. As usual, we assume that $\mathbf{A} = \{\mathbf{a}_{ij}\}$ is an $m \times n$ data matrix where $\mathbf{a}_{ij}$ is the i-th observation of the j-th predictor, $i = 1, ..., m$ and $j = 1, ..., n$. Instead of a vector $\mathbf{y}$ we now have an $m \times k$ matrix $\mathbf{Y} = \{\mathbf{y}_{il}\}$ of observed response variables, where each column-vector $\mathbf{y}^l$ is an m-dimensional sample of the l-th response variable, $l = 1, ..., k$. The goal is to learn simultaneously the $n \times k$ matrix of coefficients $\mathbf{X} = \{\mathbf{x}_{jl}\}$ for k linear regression models, one per each response variable. Here $\mathbf{x}_j l$ denotes the coefficient of the j-th predictor in the l-th model. We will denote by $\mathbf{x}_j$ the j-th row in $\mathbf{X}$ (i.e., the coefficients of the j-th predictor across all regression tasks), and by $\mathbf{y}^l$ and $\mathbf{x}^l$ the l-th columns in the matrices $\mathbf{Y}$ and $\mathbf{X}$, respectively.

The optimization problem for the *simultaneous variable selection* with multiple responses is stated as follows (Turlach et al., 2005):

$$\min_{\mathbf{x}_{11},...\mathbf{x}_{np}} \sum_{l=1}^{k} \sum_{i=1}^{m} (\mathbf{y}_{il} - \sum_{j=1}^{n} \mathbf{a}_{ij}\mathbf{x}_{jl})^2, \tag{6.8}$$

$$\text{subject to} \quad \sum_{j=1}^{n} \max(|\mathbf{x}_{j1}|, ..., |\mathbf{x}_{jk}|) \leq t. \tag{6.9}$$

This formulation can be also rewritten using the corresponding Lagrangian form, and the vector notation, as follows:

$$\min_{\mathbf{X}} \sum_{l=1}^{k} ||\mathbf{y}^l - \mathbf{A}\mathbf{x}^l||_2^2 + \lambda \sum_{j=1}^{n} ||\mathbf{x}_j||_\infty, \tag{6.10}$$

where
$$||\mathbf{x}_j||_\infty = \max\{|\mathbf{x}_1^j|, ..., |\mathbf{x}_n^j|\}$$

is called the l_∞ norm. This regularization penalty is called the *block l_1/l_∞-norm*, since it applies the l_1-norm (the sum) to the "blocks" (groups) of coefficients, where each group is a row of $\mathbf{X}$ (i.e., it corresponds to a predictor variable), while the l_∞-norm is used within each row (i.e., across all tasks). Similarly to the l_1/l_2 block norm, the l_1 part encourages sparsity across the groups, i.e. the rows of $\mathbf{X}$, while the l_∞ penalty only depends on the largest coefficient within each row, and thus no extra penalty is incurred by making all coefficients nonzero, or even the same as the maximum. Thus, no sparsity is enforced within the group, and if a predictor (row) is selected, it enters all k regression models with nonzero coefficients. Thus, both l_1/l_∞ and l_1/l_2 block norms have similar effect, enforcing sparsity across the groups, but not within them.

6.5   Generalizations

6.5.1   Block l_1/l_q-Norms and Beyond

The l_1/l_2 and l_1/l_∞ penalties discussed above are the two most commonly used members in a large family of the *block l_1/l_q-norm* penalties, defined as

$$\sum_{j=1}^{J} c_j ||\mathbf{x}_{G_j}||_q, \tag{6.11}$$

where $q > 1$, G is the partition of variables into (non-intersecting) groups, and the weight c_j accounts for varying group sizes, just as it was defined for group LASSO.

An even more general class of penalties, called the *composite absolute penalties (CAP)*, was suggested by (Zhao et al., 2009). A set of groups of predictors $G = \{G_1, ..., G_J\}$ is assumed to be given, where the groups are either non-overlapping, as in the group LASSO, or otherwise forming a hierarchical structure of overlapping groups. A set of norm parameters $q = (q_0, q_1, ..., q_J)$, $q_i > 0$, $i \in \{0, 1, ..., J\}$ is also given. Let

$$Z_j = ||\mathbf{x}_{G_j}||_{q_j}$$

denote the q_j-norm of $\mathbf{x}_{G_j}$ vector over the j-th group of variables. Then the general CAP penalty is defined as the q_0-norm of the vector $Z = (Z_1, ..., Z_J)$, to the q_0-th power:

$$||Z||_{q_0}^{q_0} = \sum_{j=1}^{J} c_j |Z_j|^{q_0}. \tag{6.12}$$

Here the q_0-norm determines the relation among the groups, while the q_i-norms determine within-group relations among the variables in each group. As discussed above, $q_j > 1$ results in all nonzero coefficients within the group G_j, while $q_0 = 1$ yields l_1-norm over the groups, and thus, combined with $q_j > 1$, promotes sparsity at the group level. In summary, CAP penalties generalize both group LASSO and simultaneous LASSO by including l_2 and l_∞ group norms as particular cases; moreover, they allow for overlapping groups and hierarchical variable selection. (Zhao et al., 2009) propose several algorithms for learning CAP-regularized model parameters: in general cases, they use BLASSO algorithm (Zhao and Yu, 2007) that approximates the regularization path and works by combining forward and backward steps, such as adding or deleting a variable. In the specific case of the usual sum-squared loss and the l_∞ within-group penalty, combined with l_1 across-group penalty, they introduce exact path-following algorithms similar to LARS (homotopy), called iCAP (for non-overlapping groups) and hiCAP (for hierarchical variable selection).

6.5.2 Overlapping Groups

While the original group LASSO formulation assumes non-overlapping groups, there are multiple recent extensions of this approach to more complex, overlapping group structures; see (Jenatton et al., 2011a; Jenatton, 2011) for more details. The block l_1/l_q penalty with overlapping groups results in a sparsity pattern that is much less obvious; in order to obtain a desirable nonzero pattern (also called the solution's support), the set of overlapping groups must be carefully designed. Recall that the coefficients of the variables belonging to any given group are set to zero simultaneously; thus, unions of the groups correspond to possible zero patterns, and, consequently, possible supports (nonzero patterns) correspond to the *intersections* of groups' *complements*. For example, let us consider a set of overlapping groups in Figure 6.8, over a set of variables with some linear ordering. It is easy to see that we can get all possible contiguous supports by setting to zero various combinations of groups; contiguous supports are naturally occurring in a variety of practical applications that

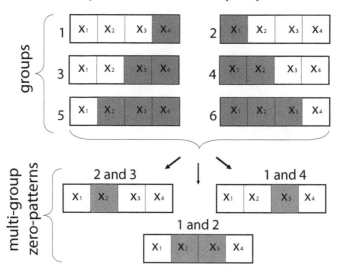

FIGURE 6.8: A set of overlapping groups (shaded) over a sequence of linearly ordered predictor variables; zero-pattern (white) and support (shaded) of solutions resulting from setting some of the groups to zero.

involve time series. Another type of sequential data, as discussed, for example, in (Jenatton et al., 2011a), involves CGH arrays used for tumor diagnosis.

More general classes of support patterns include two-dimensional convex support, two-dimensional block structures, and hierarchical group structures; again, see (Jenatton et al., 2011a; Jenatton, 2011) for more details. Hierarchical group structure assumes that the variables are organized in a tree or a forest, and setting a variable to zero in the model implies that all its descendant variables are also zeros (or, in other words, a variable participates in the support, or nonzero pattern, only if all its ancestors are in the support). For example, Figure 6.9 shows a a hierarchical set of groups over a tree containing nine variables, and a sparsity pattern and support resulting from setting some of the groups to zero. As it was mentioned before, hierarchical groups were considered first by (Zhao et al., 2009). Another common example of hierarchical grouping is the sparse group LASSO (Friedman et al., 2010) that assumes a set of multivariate non-overlapping groups (a forest), and a set of groups formed by all leaves; thus, there is a variable selection both at the level of non-overlapping groups, as in the regular group LASSO, and at the variable level within each group, hence the name "sparse group" LASSO. Hierarchical groups are widely used in a variety of applications, including computational biology (Kim and Xing, 2010), fMRI analysis (Jenatton et al., 2011b), topic modeling and image restoration (Jenatton et al., 2011c), and many others.

A further study of group LASSO with overlapping groups was presented in (Jacob et al., 2009) and (Obozinski et al., 2011). Unlike previous examples where the supports produced by overlapping groups were intersection-closed (i.e., intersection of possible supports still belongs to the set of possible supports produced by a given

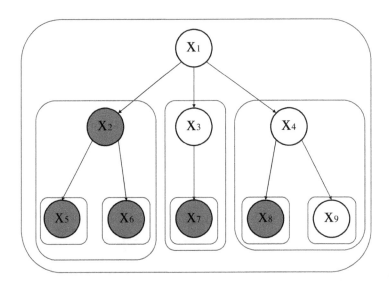

FIGURE 6.9: A hierarchical set of overlapping groups (shown within contours) over a set of predictor variables organized in a tree; zero-pattern (shaded) and support (white) of solutions resulting from setting to zero groups (x_2, x_5, x_6), (x_7), and (x_8).

set of groups), (Jacob et al., 2009; Obozinski et al., 2011) consider the case of union-closed supports. This formulation is referred to as "latent group LASSO" since it is using group LASSO penalty on a set of latent, or hidden, variables.

6.6 Applications

6.6.1 Temporal Causal Modeling

One natural application of the group LASSO approach is learning sparse vector-autoregressive models, where the groups correspond to time-lagged variables of the same time-series. Based on this idea, (Lozano et al., 2009a) propose a grouped graphical Granger modeling method for gene expression regulatory network discovery. More specifically, given a time-series of observations for each predictor variable, $x_i^1,...,x_i^t$, and the target variable, $y^1,...,y^t$, we may want to learn a regression model predicting the target variable y^t at time t from all previous observations with time lag up to k. For each i, the set of observations $\{x_i^{t-k}, ..., x_i^{t-1}, x_i^{t-1}\}$ can be naturally considered as a group of variables that need to be simultaneously included into our sparse model if, for example, the predictor X_i has a causal effect on the target variable, or simultaneously excluded from the model if there is no such effect.

6.6.2 Generalized Additive Models

As it was already mentioned, some natural grouping of variables arises in many applications. For example, (Bakin, 1999) used the group LASSO approach for learning *generalized additive models (GAMs)*. GAMs generalize the conventional linear regression model as follows:

$$g(E(Y)) = \sum_i f_i(x_i), \quad f_i(\mathbf{x}) = \sum_{j=1}^{J} \alpha_{ij} h_k(\mathbf{x}),$$

where $E(Y)$ is the expectation of the response variable Y, and $h_k(\mathbf{x})$ are the basis functions in the basis expansion of each component function $f_i(\mathbf{x})$. Note that when $f_i(\mathbf{x})$ are linear functions, and $g(z) = z$ (identity function), then the model reduces to simple linear regression. In (Bakin, 1999), the group LASSO type of penalty was imposed on the parameters α_{ij}, where each group includes the coefficients for the same component function f_i, i.e. $g_i = \{\alpha_{ij} : j = 1, ...J\}$; such parameter grouping corresponds to imposing sparsity over the component functions, i.e. choosing a subset of them to be included in the GAM model.

6.6.3 Multiple Kernel Learning

The group LASSO approach was recently used in the multiple kernel learning framework discussed in (Lanckriet et al., 2004; Bach et al., 2004; Bach, 2008b). First, we will briefly discuss the standard single kernel learning; for more details, see, for example, (Shawe-Taylor and Cristianini, 2004). The objective here is to learn a potentially *nonlinear* model that first maps the input variables (predictors) into some high-dimensional (possibly infinite-dimensional) features, and then uses a linear mapping from the features to the output variable (target). This can be done efficiently using the so-called *kernel trick*, which avoids an explicit computation of the mapping from the input to the feature space, and only involves computing a *kernel* $\mathbf{K}(\mathbf{x}_1, \mathbf{x}_2)$ as discussed below.

More specifically, given a set of m data points $(\mathbf{x}^1, y^1), ..., (\mathbf{x}^m, y^m)$, where $\mathbf{x}^i$ is the i-th observation of the input vector $\mathbf{x}$, and y^i is the corresponding observation of the output variable, the traditional single kernel learning aims at constructing a predictor

$$y = \langle \mathbf{w}, \Phi(\mathbf{x}) \rangle,$$

where $\Phi : X \to F$ is a function from the input space X into the *feature space* F. The feature space here is a *reproducing kernel Hilbert space* associated with the *kernel function* $\psi(\cdot, \cdot) : X \times X \to R$, and $\langle \cdot, \cdot \rangle : F \times F \to R$ is the inner product function in that space. In general, F is an infinite-dimensional function space, though for simplicity, we focus here on a finite-dimensional case when $\Phi(\mathbf{x}) \in R^p$, and thus $\langle \mathbf{w}, \Phi(\mathbf{x}) \rangle = \mathbf{w}^T \Phi(\mathbf{x})$. Note that the above predictor is linear in features $\Phi(\mathbf{x})$ but nonlinear in the original input $\mathbf{x}$ when the mapping $\Phi(\cdot)$ is nonlinear.

The parameters of the predictor are estimated by solving the optimization problem:

$$\min_{\mathbf{w} \in F} \sum_{i=1}^{m} L(y^i, \mathbf{w}^T \Phi(\mathbf{x}^i)) + \lambda \|\mathbf{w}_i\|_2^2,$$

where L is some loss function, such as, for example, the quadratic loss. According to the representer theorem (Kimeldorf and Wahba, 1971), the solution to the above problem has the form $\mathbf{w} = \sum_{j=1}^{m} \alpha_j \Phi(\mathbf{x}^j)$. Using this expression, and denoting by α the column vector $(\alpha_1, ..., \alpha_m)^T$, and by $(\mathbf{K}\alpha)_i$ the i-th element of the corresponding product of a matrix $\mathbf{K}$ and a vector α, we can rewrite the optimization problem as

$$\min_{\alpha \in R^m} \sum_{i=1}^{m} L(y^i, (\mathbf{K}\alpha)_i) + \lambda \alpha^T \mathbf{K}\alpha,$$

where $\mathbf{K}$ is the kernel matrix defined as

$$\mathbf{K}_{ij} = \psi(\mathbf{x}^i, \mathbf{x}^j) = \langle \Phi(\mathbf{x}^i), \Phi(\mathbf{x}^j) \rangle.$$

The above optimization problem can be solved efficiently provided that the kernel matrix is easy to compute; some examples include linear, polynomial, and Gaussian kernels. For more information on kernel-based learning we refer the reader to (Shawe-Taylor and Cristianini, 2004). There are also several tutorials available online.

The multiple kernel learning (Lanckriet et al., 2004; Bach et al., 2004; Bach, 2008b) is an extension of the standard kernel-based learning to the case when the kernel can be represented as a sum of s multiple kernels, i.e.

$$\mathbf{K}(\mathbf{x}, \mathbf{x}') = \sum_{i=1}^{s} \alpha_i K_i(\mathbf{x}, \mathbf{x}').$$

A multiple kernel situation naturally arises in some applications, such as, for example, fusion of the multiple heterogeneous data sources (Lanckriet et al., 2004). The predictor in multiple kernel cases takes the following form

$$\sum_{i=1}^{s} \mathbf{w}_i^T \Phi_i(\mathbf{x}),$$

where $\Phi_i : X \to F_i$ is the i-th feature map from the input space to the i-th feature space F_i, associated with the i-th kernel, and $\mathbf{w} \in F_i$. For simplicity, let us assume again the finite-dimensional case, i.e. $\Phi_i(\mathbf{x}) \in R^{n_i}$, though the framework also extends to infinite-dimensional spaces (Bach et al., 2004; Bach, 2008b). Using the group penalty

$$\sum_{i=1}^{s} \|\mathbf{w}_i\|_2,$$

where the i-th group is associated with the i-th kernel, allows to select in a data-driven way the best subset of feature spaces (and associated kernels), out of a potentially large amount of possible spaces (kernels).

6.6.4 Multi-Task Learning

Simultaneous variable selection with multiple responses presented in (Turlach et al., 2005) is a particular case of a general problem in machine learning, referred to as *multi-task learning*, where the objective is to simultaneously learn multiple predictive models over the same set of predictors. In general, the design matrices can be different for different tasks, or coincide as in the case of simultaneous variable selection in (Turlach et al., 2005). For example, (Liu et al., 2009a) apply the multi-task approach to learning sparse models for predicting brain activation in fMRI in response to words as stimuli. More specifically, they simultaneously learn sparse linear models to predict 20,000 voxel activations, while using the l_1/l_∞ regularization to extract a common group of predictive features out of 50,000-dimensional feature vectors that encode word meaning through words' co-occurrences with other words.

Besides the linear regression setting, multi-task learning is often applied to classification problems where the target variable is discrete, e.g., binary. In these cases, an appropriate loss function, such as logistic loss (Obozinski et al., 2010) or hinge-loss (Quattoni et al., 2009), replaces the sum-squared loss. Groups are formed by coefficients of the same predictor variable across all models (i.e., across different learning tasks).

For example, (Obozinski et al., 2010) consider the optical character recognition (OCR) problem for multiple writers, where pixel-level or stroke-level representations of the handwritten characters correspond to the features/predictors, and the task is to classify a given character into one of a specified class (e.g., a particular digit or letter). Figure 6.10a shows examples of the letter "a" written by different people, and Figure 6.10b shows the strokes extracted from the data. There are typically a few thousands of pixel- and/or stroke-level features, and only a relatively small subset of them turns out to be useful for character recognition. Thus, learning a sparse classification model looks like a natural approach. (Obozinski et al., 2010) build binary classifiers that discriminate between pairs of letters, using logistic loss with the l_1-norm regularization as the baseline. In the presence of multiple datasets, one for each individual, (Obozinski et al., 2010) aim at learning a single sparse model that shares predictors across different writers. The models are trained jointly on the data for all participating individuals, using the block l_1/l_2-norm constraint in order to select common predictive features across different models (prediction tasks). As demonstrated in (Obozinski et al., 2010), models learned by such multi-task approach are more accurate than the models learned separately for each writer, perhaps due to the fact that the image features important for character recognition are indeed shared across different subjects despite the differences in their writing styles. Note also that the number of samples for each individual writer is quite limited (e.g., between 4 and 30 samples), and thus combining the datasets with similar properties tends to improve the model's accuracy by increasing the training set size.

Another multi-task problem also discussed in (Obozinski et al., 2010) is a DNA microarray analysis problem. There are typically several thousands features corresponding to the gene expression levels, and the task is to predict a certain phenotypic property, e.g. a type of a skin cancer that patient has. A common assumption is that

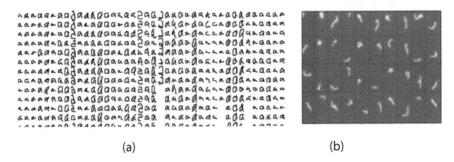

(a)   (b)

FIGURE 6.10: Sample data for the optical character recognition (OCR) problem: (a) letter "a" written by 40 different people; (b) stroke features extracted from the data.

only a limited subset of genes is relevant to the given phenotypic variable, such as presence of a particular disease, and thus a sparse model is a tool of choice. When the phenotype variables are related, such as, for example, in the case of a related set of cancers, it can be advantageous to use a multi-task approach and learn multiple sparse models sharing a common subset of variables/genes, as demonstrated by the experiments presented in (Obozinski et al., 2010).

6.7 Summary and Bibliographical Notes

In this chapter, we introduced several recently developed sparse regression techniques that extend and improve the standard LASSO approach by incorporating additional, domain-specific solution structure, besides the trivial sparsity assumption. Such methods are based on combining the l_1-norm penalty with other l_q-norms. For example, a convex combination of the l_1- and the l_2-norm penalties in the Elastic Net regression (Zou and Hastie, 2005) allows for both sparsity and *grouping*, i.e. joint inclusion or exclusion, of highly correlated variables, which is particularly relevant in applications such as brain imaging or gene microarray analysis. The fused LASSO (Tibshirani et al., 2005) imposes smoothness of regression coefficients along a specified variable ordering – an important feature in time series data. In general, when the data possess some natural variable grouping, the group LASSO (Yuan and Lin, 2006) allows for sparsity at the group level; block l_1/l_q-penalties, with both non-overlapping and overlapping groups, are a popular tool for imposing different group structures in a wide range of applications, including multi-task learning, multiple kernel learning, and various applications in signal processing, biology, and beyond. Efficient group LASSO algorithms developed in the past several years include, among others, block-coordinate descent methods (Yuan and Lin, 2006; Liu et al., 2009a), projected gradient (Quattoni et al., 2009), active set approaches (Roth and Fischer, 2008; Obozinski et al., 2010), Nesterov's method (Liu et al., 2009b), as well as greedy techniques such as group orthogonal matching pursuit (Lozano et al., 2009b). Asymptotic consistency of the group LASSO is the focus of several recent papers, including (Bach, 2008b; Liu and Zhang, 2009; Nardi and Rinaldo, 2008). Further extensions include group LASSO for logistic regression (Meier et al., 2008) and other generalized linear models (GLMs) (Roth and Fischer, 2008). Also, (Jenatton, 2011) provides a comprehensive survey of structured sparsity approaches and methods.

Chapter 7

Beyond LASSO: Other Loss Functions

In the previous chapter we discussed extensions of the basic LASSO problem to a wider class of regularizers. We will now focus on a more general class of loss functions, as discussed earlier in chapter 2, namely, on the *exponential-family* negative log-likelihood losses. Note, however, that several other popular types of loss functions remain out of scope of this book; see the bibliography section of chapter 2 for a brief summary.

Recall that, from a probabilistic point of view, the LASSO problem is equivalent to finding model parameters maximizing a posteriory probability (MAP) under the assumption of linear Gaussian measurements and model parameters following the Laplace prior. However, in many practical applications, a non-Gaussian model of measurement noise may be more appropriate. For example, when the measurements are discrete variables (e.g., binary or categorical, in general), such as failures in a distributed computer system (Rish et al., 2005; Zheng et al., 2005), word counts in a document, and so on, then the Bernoulli or multinomial distributions must be used instead. On the other hand, when actual transaction response times are used to infer possible performance bottlenecks in a system, exponential distribution is better suited than Gaussian, since the response time is nonnegative (Chandalia and Rish, 2007; Beygelzimer et al., 2007). Non-Gaussian observations, including binary, discrete, nonnegative, etc., variables, are also common in applications such as computational biology and medical imaging: for example, a binary variable describing the presence or absence of a particular disease given gene microarray data, or multinomial variables corresponding to different levels of emotions such as anger, anxiety, or happiness that we may want to predict based on brain images (Mitchell et al., 2004; Carroll et al., 2009). In general, classification problems with discrete class labels are one of the central topics in modern machine learning.

In this chapter, we will consider a general class of noise distributions, known as *exponential family* (Mccullagh and Nelder, 1989), which includes, besides the Gaussian distribution, many others commonly used distributions, such as Bernoulli, multinomial, exponential, gamma, chi-square, beta, Weibull, Dirichlet, and Poisson, just to name a few. The problem of recovering an unobserved vector $\mathbf{x}$ from the vector $\mathbf{y}$ of linear measurements $A\mathbf{x}$ contaminated by an exponential-family noise is known as *Generalized Linear Model (GLM)* regression. Adding the l_1-norm constraint to GLM regression allows for an efficient method of sparse signal recovery, and is often used in statistical literature (Park and Hastie, 2007).

A natural question to ask next is whether a sparse signal can be accurately recovered from linear measurements corrupted by an exponential-family noise. It turns out that this question can be answered positively, as it was shown by (Rish and Grabarnik, 2009), where the classical results of (Candès et al., 2006b) were extended to the exponential-noise case. This chapter provides a summary of those results, as well as a brief review of the recent theoretical results that address properties of general *regularized M-estimators* (maximum-likelihood estimators), including the LASSO and the l_1-regularized GLM problems as particular cases (Negahban et al., 2009, 2012).

7.1 Sparse Recovery from Noisy Observations

Let us first recall some basic results about the sparse signal recovery from noisy measurements; also, see chapters 3 and 4. We assume that $\mathbf{x}^0 \in R^n$ is a k-sparse signal, i.e. a signal with no more than k nonzero entries, where $k << n$. Let $\mathbf{A}$ be an m by n matrix that produces a vector of linear projections $\mathbf{y}^0 = A\mathbf{x}^0$, where $m << n$, and let $\mathbf{y}$ be a vector of m noisy measurements that follow some noise distribution $P(\mathbf{y}|A\mathbf{x}^0)$. Following (Candès et al., 2006b), we will assume herein that the matrix $\mathbf{A}$ satisfies the restricted isometry property (RIP) at the sparsity level k (recall that RIP was defined earlier in Chapter 3), which essentially states that every subset of columns of $\mathbf{A}$ with cardinality less than k behaves like a nearly orthonormal system.

The question we focus on here is whether the true signal $\mathbf{x}^0$ can be accurately recovered from $\mathbf{y}$, given that the measurement noise is sufficiently small. This question has been answered in the compressed sensing literature for the particular case when the noise distribution is Gaussian. Indeed, as it was shown in (Candès et al., 2006b), if (1) $||\mathbf{y} - A\mathbf{x}^0||_{l_2} \leq \epsilon$ (i.e., the noise is small), (2) $\mathbf{x}^0$ is sufficiently sparse, and the (3) matrix $\mathbf{A}$ obeys the RIP with some appropriate RIP constants, then the solution to the following l_1-norm minimization problem:

$$\mathbf{x}^* = \arg\min_{\mathbf{x}} ||\mathbf{x}||_1 \text{ subject to } ||\mathbf{y} - A\mathbf{x}||_2 \leq \epsilon \qquad (7.1)$$

approximates the true signal well. More formally, Theorem 1 in (Candès et al., 2006b) states:

Theorem 7.1. *(Candès et al., 2006b) Let S be such that $\delta_{3S} + 3\delta_{4S} < 2$, where δ_S is the S-restricted isometry constant of the matrix $\mathbf{A}$, as defined above. Then for any* *signal* $\mathbf{x}^0$ *with the support* $T^0 = \{t : x^0 \neq 0\}$, *where* $|T^0| \leq S$ *and any noise vector (perturbation)* e *with* $||e||_2 \leq \epsilon$, *the solution* $\mathbf{x}^*$ *to the problem in eq. 7.1 obeys*

$$||\mathbf{x}^* - \mathbf{x}^0||_2 \leq C_S \cdot \epsilon, \qquad (7.2)$$

where the constant C_S *may only depend on* δ_{4S}. *For reasonable values of* δ_{4S}, C_S *is well-behaved; e.g.,* $C_S \approx 8.82$ *for* $\delta_{4S} = 1/5$ *and* $C_S \approx 10.47$ *for* $\delta_{4S} = 1/4$.

Moreover, (Candès et al., 2006b) show that (1) no other recovery method "can perform fundamentally better for arbitrary perturbations of size ϵ, i.e. even if an oracle would make the actual support T^0 of $\mathbf{x}^0$ available to us, making the problem well-posed, the least-squares solution $\hat{\mathbf{x}}$ (i.e., the maximum-likelihood solution which is optimal in the absence of any other information) would approximate the true signal $\mathbf{x}^0$ with the error proportional to ϵ".

Finally, (Candès et al., 2006b) extend their result from sparse to approximately sparse vectors:

Theorem 7.2. *(Candès et al., 2006b) Let* $\mathbf{x}^0 \in R^n$ *be an arbitrary vector, and let* $\mathbf{x}_S^0$ *be the truncated vector corresponding to the S largest (absolute) values of* $\mathbf{x}^0$. *Under the assumptions of Theorem 7.1, the solution* $\mathbf{x}^*$ *to the problem in eq. 7.1 obeys*

$$||\mathbf{x}^* - \mathbf{x}^0||_2 \leq C_{1,S} \cdot \epsilon + C_{2,S} \cdot \frac{||\mathbf{x}^0 - \mathbf{x}_S^0||_1}{\sqrt{S}}. \qquad (7.3)$$

For reasonable values of δ_{4S} *the constants above are well-behaved; e.g.,* $C_{1,S} \approx 12.04$ *and* $C_{2,S} \approx 8.77$ *for* $\delta_{4S} = 1/5$.

In the rest of this chapter, we will discuss an extension of the above compressed-sensing results to the case of general exponential-family noise distributions, as it was presented in (Rish and Grabarnik, 2009).

7.2 Exponential Family, GLMs, and Bregman Divergences

Note that the $||\mathbf{y} - \mathbf{Ax}||_2 \leq \epsilon$ constraint results from the negative log-likelihood of a Gaussian variable $\mathbf{y} \sim N(\mu, \Sigma)$ with $\mu = \mathbf{Ax}$ and $\Sigma = I$ (i.e., independent unit-variance Gaussian noise):

$$-\log p(\mathbf{y}|\mathbf{Ax}^0) = f(\mathbf{y}) + \frac{1}{2}||\mathbf{y} - \mathbf{Ax}||_2^2. \qquad (7.4)$$

The Gaussian distribution is a particular member of the *exponential family* of distributions.

7.2.1   Exponential Family

Definition 8. *An **exponential family** is a parametric family of probability distributions with the probability density defined as follows[1]:*

$$p_{\psi,\theta}(\mathbf{y}) = p_0(\mathbf{y})e^{\theta^T T(\mathbf{y}) - \psi(\theta)}, \tag{7.5}$$

*where θ is called the **natural parameter** of the distribution. $T(\mathbf{x})$ is a vector of **sufficient statistics** that fully summarizes the data $\mathbf{y}$ within the density function, i.e., the value of the density function is the same for any two data vectors $\mathbf{y}_1$ and $\mathbf{y}_2$, as long as $T(\mathbf{y}_1)$ and $T(\mathbf{y}_2)$. The function $\psi(\theta)$ is strictly convex and differentiable, and is called the **cumulant function**, or the **log-partition function**; this function uniquely defines a particular member distribution of the exponential family, and can be computed as:*

$$\psi(\theta) = \log \int p_0(\mathbf{y})e^{\theta^T T(\mathbf{y})} d\mathbf{y}. \tag{7.6}$$

*Finally, $p_0(\mathbf{y})$ is a nonnegative function, called the **base measure**, which only depends on the data $\mathbf{y}$, and is independent of the parameter θ.*

We now demonstrate how two commonly used distributions, the Gaussian distribution and the Bernoulli distribution, can be written in the exponential-family form.

Gaussian distribution, unknown mean, and unknown variance. The univariate Gaussian distribution with some mean μ and standard deviation σ can be written as

$$
\begin{aligned}
p(y) &= \frac{1}{\sqrt{2\pi\sigma^2}} exp\{-\frac{(y-\mu)^2}{2\sigma^2}\} = \\
&= \frac{1}{\sqrt{2\pi}} exp\{\frac{\mu}{\sigma^2}y - \frac{1}{2\sigma^2}y^2 - \frac{1}{2\sigma^2}\mu^2 - \log\sigma\},
\end{aligned} \tag{7.7}
$$

and thus is a member of the exponential family defined in eq. 7.5, where

$$
\begin{aligned}
T(y) &= (y, y^2)^T, \\
\theta &= (\mu/\sigma^2, -1/(2\sigma^2))^T, \\
\psi(\theta) &= \mu^2/(2\sigma^2) + \log\sigma = -\theta_1^2/(4\theta_2) - 0.5\log(-2\theta_2), \\
p_0(y) &= 1/\sqrt{2\pi}.
\end{aligned} \tag{7.8}
$$

Gaussian distribution, unknown mean, unit variance. In a specific case of the univariate Gaussian distribution with known variance, and particularly, unit variance, i.e. $\sigma = 1$, we can simplify the above expressions as follows:

$$
\begin{aligned}
p(y) &= \frac{1}{\sqrt{2\pi}} exp\{-\frac{(y-\mu)^2}{2}\} = \\
&= \frac{exp\{-y^2/2\}}{\sqrt{2\pi}} exp\{\mu y - \mu^2/2\},
\end{aligned} \tag{7.9}
$$

---

[1]Note that, for the simplicity of exposition, we focus here on the exponential family in a *canonical form*; in general, $\theta^T T(\mathbf{y})$ must be replaced by $\eta(\theta)^T T(\mathbf{y})$, where η is some known function; it is always possible, however, to convert an exponential family to the canonical form.

and thus we get:

$$T(y) = y, \tag{7.10}$$
$$\theta = \mu,$$
$$\psi(\theta) = \mu^2/2 = \theta^2/2,$$
$$p_0(y) = exp\{-y^2/2\}/\sqrt{2\pi}.$$

Another commonly used member of the exponential family is the Bernoulli distribution, i.e. the distribution of a binary random variable that takes values 0 and 1, with a single parameter $q = p(y = 1)$. The Bernoulli distribution can be written as follows:

$$p(y) = q^y(1-q)^{1-y} = exp\{\log(q^y(1-q)^{1-y})\} = \tag{7.11}$$
$$= exp\{y \log q + (1-y)\log(1-q)\} =$$
$$= exp\{y \log \frac{q}{1-q} + \log(1-q)\} =$$
$$= exp\{y\theta - \log(1+e^\theta)\},$$

namely, we obtained an exponential-family distribution where

$$T(y) = y, \tag{7.12}$$
$$\theta = \log \frac{q}{1-q},$$
$$\psi(\theta) = \log(1+e^\theta),$$
$$p_0(y) = 1.$$

Note that we can also rewrite the definition of the exponential family distributions in eq. 7.5 as a function of $T(\mathbf{y})$, rather than a function of $\mathbf{y}$, since, by definition of the sufficient statistic, $p_0(\mathbf{y})$ can be written as some base measure $p_0'(T(\mathbf{y}))$. Thus, from now on, we will define exponential family simply as

$$p_{\psi,\theta}(\mathbf{y}) = p_0(\mathbf{y})e^{\theta^T \mathbf{y} - \psi(\theta)}, \tag{7.13}$$

assuming that $\mathbf{y}$ is a sufficient statistic; note that this version of the exponential family definition is often used in the literature; see, for example, (Collins et al., 2001; Sajama and Orlitsky, 2004; Banerjee et al., 2004; Rish et al., 2008; Li and Tao, 2010), just to name a few.

7.2.2 Generalized Linear Models (GLMs)

Recall that the standard linear regression model, $y = \sum a_i x_i + \epsilon = \mathbf{a}^T \mathbf{x} + \epsilon$, also called the ordinary least-squares (OLS), assumes that the mean of the observed (output) random variable y is linear in the input vector $\mathbf{a}$ of random variables, i.e. $E(y) = \mathbf{a}^T \mathbf{x}$, where $\mathbf{x}$ are the model parameters, and that the noise ϵ is zero-mean

Gaussian noise, which implies that the distribution of $\mathbf{y}$ is Gaussian with the mean $\mu = \mathbf{a}^T \mathbf{x}$. Assuming that the variance of the noise is known (e.g., unit variance after normalizing the data), the maximum-likelihood parameter estimation of the linear regression model, given a vector of observations $\mathbf{y}$, and a matrix of (row-vector) inputs $\mathbf{A}$, consists in minimizing the sum-squared loss $||\mathbf{y} - \mathbf{A}\mathbf{x}||^2$, as shown in eq. 7.4.

A generalization of the linear regression model to the exponential-family noise is known as the *Generalized Linear Model (GLM)*[2]. GLM assumes that the mean $\mu = E(y)$ is a (generally, nonlinear) function of the *linear predictor* $\theta = \mathbf{a}^T \mathbf{x}$, which corresponds to the natural parameter of the exponential-family distribution describing the observation noise, i.e.

$$E(y) = \mu(\theta) = f^{-1}(\mathbf{a}^T \mathbf{x}), \tag{7.14}$$

where the function f is called the *link function*, as it "links" the mean and the linear predictor, namely, $\theta = f(\mu)$ and $\mu = f^{-1}(\theta)$. It can be shown that

$$\mu = E(y) = f^{-1}(\theta) = \nabla\psi(\theta). \tag{7.15}$$

In case of the unit-variance Gaussian, $\theta = \mu$, as shown above in eq. 7.11, and thus the link function $f(\mu)$ and its inverse $f^{-1}(\theta)$ are both simply *identity functions*, i.e. $f(\mu) = \mu$, which yields the standard *linear regression model*:

$$E(y) = \mathbf{a}^T \mathbf{x},$$

while for the Bernoulli noise with the parameter $q = p(y = 1)$, the mean is $\mu = q$, and thus the link function is the *logit* function $f(\mu) = \log\frac{\mu}{1-\mu}$ (see eq. 7.13), which is the inverse of the *logistic function* $f^{-1}(\theta) = \frac{1}{1+exp^{-\theta}}$, and yields the *logistic regression model*:

$$E(y) = \frac{1}{1 + e^{-\mathbf{a}^T \mathbf{x}}}.$$

7.2.3 Bregman Divergence

As shown by (Banerjee et al., 2004), there is a bijection between the exponential-family densities $p_{\psi,\theta}(\mathbf{y})$ and the so-called *Bregman divergences* $d_\phi(\mathbf{y}, \mu)$, defined below, so that each exponential-family density can be also expressed as

$$p_{\psi,\theta}(\mathbf{y}) = exp(-d_\phi(\mathbf{y}, \mu))f_\phi(\mathbf{y}), \tag{7.16}$$

where $\mu = \mu(\theta) = E_{p_{\psi,\theta}}(y)$ is the *expectation parameter* corresponding to θ, as discussed in the previous section, ϕ is the (strictly convex and differentiable) Legendre conjugate of ψ, $f_\phi(\mathbf{y})$ is a uniquely determined function, and $d_\phi(\mathbf{y}, \mu)$ is the corresponding Bregman divergence defined as follows.

---

[2]It is important not to confuse it with the *general linear model*, which is also abbreviated as GLM; this model is still linear, though the output is now multivariate, i.e. y is an m-dimensional vector of random variables, rather than a single variable, and, given n samples, general linear model is written as $\mathbf{Y} = \mathbf{A}\mathbf{X} + \mathbf{U}$, where Y is a matrix with series of multivariate measurements, X is a design matrix, B is a matrix of parameters, and U is a noise matrix.

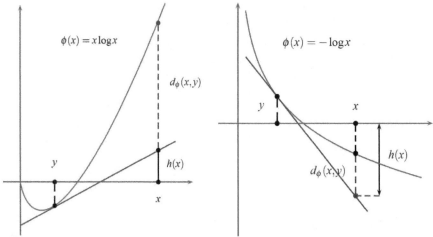

(a) Relative entropy (KL-divergence) (b) Itakura-Saito distance (Burg divergence)

FIGURE 7.1: Examples of Bregman divergences.

Definition 9. *Given a strictly convex function* $\phi : S \to \mathbb{R}$ *defined on a convex set* $S \subseteq \mathbb{R}$, *and differentiable on the interior of S, $int(S)$ (Rockafeller, 1970), the* **Bregman divergence** $d_\phi : S \times int(S) \to [0, \infty)$ *is defined as*

$$d_\phi(\mathbf{x}, \mathbf{y}) = \phi(\mathbf{x}) - \phi(\mathbf{y}) - (\mathbf{x} - \mathbf{y})^T \nabla \phi(\mathbf{y}), \qquad (7.17)$$

where $\nabla \phi(\mathbf{y})$ *is the gradient of* ϕ.

In other words, the Bregman divergence can be thought of as the difference between the value of ϕ at point $\mathbf{x}$ and the value of the first-order Taylor expansion of ϕ around point $\mathbf{y}$ evaluated at point $\mathbf{x}$ – see, for example, Figures 7.1a and 7.1b, where $h(x) = \phi(y) + (\mathbf{x} - \mathbf{y})^T \nabla \phi(\mathbf{y})$.

Table 7.1 (derived from Tables 1 and 2 in (Banerjee et al., 2005)) shows particular examples of commonly used exponential-family distributions and their corresponding Bregman divergences. For example, the unit-variance Gaussian distribution leads to square loss, multivariate spherical Gaussian (diagonal covariance/independent variables) gives rise to Euclidean distance, a multivariate Gaussian with the inverse-covariance (concentration) matrix C leads to Mahalanobis distance, Bernoulli distribution corresponds to logistic loss, exponential distribution leads to Itakura-Saito distance, while a multinomial distribution corresponds to the KL-divergence (relative entropy).

In summary, this section introduced three closely related concepts: exponential-family distributions, generalized linear models (GLMs), and Bregman divergences. As illustrated in Figure 7.2, there is a one-to-one mapping between each pair of these concepts. A specific exponential-family distribution is associated with a particular GLM, and vice versa; GLM extends the standard linear regression model

TABLE 7.1:   Examples of commonly used exponential-family distributions and their corresponding Bregman divergences.

Distribution/ Domain	$p_\theta(y)$	μ	$\phi(\mu)$	$d_\phi(y,\mu)$	Divergence
1D Gaussian $\mathbb{R}$	$\frac{1}{\sqrt{2\pi\sigma^2}}e^{-\frac{(x-a)^2}{2\sigma^2}}$	a	$\frac{1}{2\sigma^2}\mu^2$	$\frac{1}{2\sigma^2}(y-\mu)^2$	square loss
Bernoulli $\{0,1\}$	$q^y(1-q)^{1-y}$	q	$\mu\log\mu + (1-\mu)\log(1-\mu)$	$y\log(\frac{y}{\mu})+ (1-y)\log(\frac{1-y}{1-\mu})$	logistic loss
Exponential R_{++}	$\lambda e^{-\lambda y}$	$1/\lambda$	$-\log\mu - 1$	$\frac{y}{\mu} - \log(\frac{y}{\mu}) - 1$	Itakura-Saito
n-dim. Multinomial n-simplex	$\frac{N!}{\prod_{j=1}^n y_j!}\prod_{j=1}^n q_j^{y_j}$	$[Nq_j]_{j=1}^{n-1}$	$\sum_{j=1}^n \mu_j\log(\frac{\mu_j}{N})$	$\sum_{j=1}^n y_j\log(\frac{y_j}{\mu_j})$	KL-divergence
n-dim. Sph. Gauss. $\mathbb{R}^n$	$\frac{1}{\sqrt{(2\pi\sigma^2)^n}}e^{-\frac{||x-a||_2^2}{2\sigma^2}}$	a	$\frac{1}{2\sigma^2}||\mu||_2^2$	$\frac{1}{2\sigma^2}||y-\mu||_2^2$	squared Euclidean
n-dim. Gaussian $\mathbb{R}^n$	$\frac{\sqrt{\det(C)}}{\sqrt{(2\pi)^n}}e^{-\frac{(y-a)^TC(y-a)}{2}}$	a	$\frac{\mu^TC\mu}{2}$	$\frac{(y-\mu)^TC(y-\mu)}{2}$	Mahalanobis [3]

by assuming exponential-family, rather than just Gaussian, noise in the model, and thus a generally nonlinear relation $E(y) = f^{-1}(a^T x)$ between the linear predictor (linear function of the input variables a, with parameters x) and the mean of the output variable y. The linear predictor here corresponds to the natural parameter of the distribution $\theta = a^T x$, and the link function f relates to the mean parameter μ – and to the log-partition function of the exponential-family distribution as follows: $\mu = f^{-1}(\theta) = \nabla\psi(\theta)$.

On the other hand, as we discussed earlier, there is a bijection between the exponential-family densities $p_{\psi,\theta}(y)$ and the Bregman divergences $d_\phi(y,\mu)$, via the Legendre duality of ϕ and ψ.

Thus, fitting a GLM model is equivalent to maximum-likelihood parameter estimation with exponential-family noise assumption, and also equivalent to minimizing the corresponding Bregman divergence.

7.3   Sparse Recovery with GLM Regression

Let us now consider the following constrained l_1-norm minimization problem that generalizes the standard noisy compressed sensing problem of (Candès et al., 2006b):

$$\min_{x} ||x||_1 \text{ subject to } \sum_i d(y_i, \mu(a_i x)) \leq \epsilon, \qquad (7.18)$$

$$\log p_{\psi,\theta}(\mathbf{y}) = \mathbf{y}\theta - \psi(\theta) + \log p_0(\mathbf{y})$$

$$\theta = \mathbf{a}^T\mathbf{x}$$
$$f^{-1}(\theta) = \nabla\psi(\theta)$$

Legendre duality:

$$\psi(\theta) \Leftrightarrow \phi(\mu)$$

Generalized Linear Models

$$E(y) = \mu(\theta) = f^{-1}(\mathbf{a}^T\mathbf{x})$$

Bregman Divergences

$$d_\phi(\mathbf{x},\mathbf{y}) = \phi(\mathbf{x}) - \phi(\mathbf{y})$$
$$- (\mathbf{x} - \mathbf{y})^T\nabla\phi(\mathbf{y})$$

Fitting GLM $\Leftrightarrow$ maximizing exp-family likelihood $\Leftrightarrow$
$\Leftrightarrow$ minimizing Bregman divergence

FIGURE 7.2: One-to-one correspondences between the exponential-family distributions, Bregman divergences, and Generalized Linear Models (GLMs).

where $d(\mathbf{y}_i, \mu(\mathbf{a}_i\mathbf{x}))$ is the Bregman divergence between the noisy observation $\mathbf{y}_i$ and the mean parameter of the corresponding exponential-family distribution with the natural parameter $\theta_i = \mathbf{A}_i\mathbf{x}$. Note that using the Lagrangian form we can write the above problem as

$$\min_{\mathbf{x}} \lambda||\mathbf{x}||_1 + \sum_i d(\mathbf{y}_i, \mu(\mathbf{a}_i\mathbf{x})), \qquad (7.19)$$

where the coefficient λ is the Lagrange multiplier uniquely determined by ϵ. This problem is known as an l_1-norm regularized GLM regression, and includes as a particular case a standard l_1-norm regularized linear regression, in which case $\mu(\mathbf{a}_i\mathbf{x}) = \mathbf{a}_i\mathbf{x}$ and the Bregman divergence simply reduces to the Euclidian distance.

As shown in (Rish and Grabarnik, 2009), the result in Theorem 7.1 can be extended to the case of exponential-family noise. Specifically, if: (1) the noise is small, (2) $\mathbf{x}^0$ is sufficiently sparse, and (3) the matrix $\mathbf{A}$ obeys the restricted isometry property (RIP) with appropriate RIP constants, then the solution to the above problem approximates the true signal well.

Theorem 7.3. *Let S be such that $\delta_{3S} + 3\delta_{4S} < 2$, where δ_S is the S-restricted isometry constant of the matrix $\mathbf{A}$, as defined above, and let $\mathbf{a}_i$ denote the i-th row of $\mathbf{A}$. Then for any signal $\mathbf{x}^0$ with the support $T^0 = \{t : \mathbf{x}^0 \neq 0\}$, where $|T^0| \leq S$, and for any vector $\mathbf{y} = (y_1, ..., y_n)$ of noisy linear measurements where*

1. the noise follows exponential-family distributions $p_{\theta_i}(y_i)$, with the natural parameter $\theta_i = (\mathbf{a}_i^T\mathbf{x}^0)$,

2. the noise is sufficiently small, i.e. $\forall i,\ d_{\phi_i}(y_i, \mu(\mathbf{a}_i^T\mathbf{x}^0)) \leq \epsilon$, and

3. *each function $\phi_i(\cdot)$ (i.e., the Legendre conjugate of the corresponding log-partition function, uniquely defining the Bregman divergence), satisfies the conditions imposed by at least one of the Lemmas below,*

the solution $\mathbf{x}^*$ to the problem in eq. 7.18 obeys

$$||\mathbf{x}^* - \mathbf{x}^0||_{l_2} \leq C_S \cdot \delta(\epsilon), \tag{7.20}$$

where C_S is the constant from Theorem 7.1 (Candès et al., 2006b), and $\delta(\epsilon)$ is a continuous monotone increasing function of ϵ s.t. $\delta(0) = 0$ (and thus $\delta(\epsilon)$ is small when ϵ is small). A particular form of this function depends on particular members of the exponential family.

Proof. Following the proof of Theorem 7.1 (Candès et al., 2006b), we only need to show that the "tube constraint" (condition 1) still holds (the rest of the proof remains unchanged), i.e. that

$$||\mathbf{A}\mathbf{x}^* - \mathbf{A}\mathbf{x}^0||_{l_2} \leq \delta(\epsilon), \tag{7.21}$$

where δ is some continuous monotone increasing function of ϵ, and $\delta(0) = 0$, so its small when ϵ is small. This was a trivial consequence of the triangle inequality in the case of Euclidean distance; however, triangle inequality does not hold, in general, for Bregman divergences, and thus a different proof must be provided for the tube constraint, possibly for each type of Bregman divergence (exponential-family distribution). Since

$$||\mathbf{A}\mathbf{x}^* - \mathbf{A}\mathbf{x}^0||_{l_2}^2 = \sum_{i=1}^{m}(\mathbf{a}_i^T\mathbf{x}^* - \mathbf{a}_i^T\mathbf{x}^0)^2 = \sum_{i=1}^{m}(\theta_i^* - \theta_i^0)^2,$$

we will need to show that $|\theta_i^* - \theta_i^0| < \beta(\epsilon)$, where $\beta(\epsilon)$ is a continuous monotone increasing function of ϵ s.t. $\beta(0) = 0$ (and thus $\beta(\epsilon)$ is small when ϵ is small); then in eq. 7.21 we get $\delta(\epsilon) = \sqrt{m} \cdot \beta(\epsilon)$. Lemma 7.5 provides the proof of this fact for a class of exponential-family distributions with bounded $\phi''(y)$ (where $\phi(y)$ is the Legendre conjugate of the log-partition function that uniquely determines the distribution). However, for several members of the exponential family (e.g., Bernoulli distribution) this condition is not satisfied, and those cases must be handled individually. Thus, separate proofs are provided for several different members of the exponential family in Lemmas 7.6, 7.7, and 7.8, and particular expressions for $\beta(\epsilon)$ are obtained in each case. Note that for simplicity's sake, we only consider univariate exponential-family distributions, corresponding to the case of independent noise for each measurement y_i, which was effectively assumed in standard problem formulation that used Euclidean distance corresponding to a spherical Gaussian distribution, i.e. a vector of independent Gaussian variables. However, Lemma 7.5 below can be extended from scalar to vector case, i.e. to multivariate exponential-family distributions that do not necessarily imply independent noise. Lemma 7.8 will provide a specific case of such distribution - a multivariate Gaussian with concentration matrix C.

The "cone constraint" part of the proof in (Candès et al., 2006b) remains intact; it is easy to see that it does not depend on the particular constraint in the l_1-minimization problem 7.18, and only makes use of the sparsity of $\mathbf{x}^0$ and l_1-optimality of $\mathbf{x}^*$. Thus, we can simply substitute $\|\mathbf{A}h\|_2$ by $\delta(\epsilon)$ in eq. 13 on page 8 in the proof of Theorem 7.1 (Candès et al., 2006b), or, equivalently, replace 2ϵ (that was shown to bound $\|\mathbf{A}h\|_2$) by $\delta(\epsilon)$ in eq. 14. □

Similarly to the sparse signal case (Theorem 7.1 (Candès et al., 2006b)), the only change we have to make in the proof of Theorem 7.2 for the general case of approximable, rather than sparse, signals, relates to the tube constraint. Thus, once we showed it for Theorem 7.3 above, the generalization to approximable signals follows automatically:

Theorem 7.4. *Let $\mathbf{x}^0 \in R^m$ be an arbitrary vector, and let $\mathbf{x}^0_S$ be the truncated vector corresponding to the S largest values of $\mathbf{x}^0$ (in absolute value). Under the assumptions of Theorem 7.3, the solution $\mathbf{x}^*$ to the problem in eq. 7.18 obeys*

$$\|\mathbf{x}^* - \mathbf{x}^0\|_2 \le C_{1,S} \cdot \delta(\epsilon) + C_{2,S} \cdot \frac{\|\mathbf{x}^0 - x^0_S\|_1}{\sqrt{S}}, \qquad (7.22)$$

where $C_{1,S}$ and $C_{2,S}$ are the constants from Theorem 7.2 (Candès et al., 2006b), and $\delta(\epsilon)$ is a continuous monotone increasing function of ϵ s.t. $\delta(0) = 0$ (and thus $\delta(\epsilon)$ is small when ϵ is small). A particular form of this function depends on particular members of the exponential family.

The following lemma states the sufficient conditions for the "tube constraint" in eq. 7.21 to hold in general cases of arbitrary exponential-family noise, provided that $\phi''(y)$ exists and is bounded on the appropriate intervals.

Lemma 7.5. *Let y denote a random variable following an exponential-family distribution $p_\theta(y)$, with the natural parameter θ, and the corresponding mean parameters $\mu(\theta)$. Let $d_\phi(y, \mu(\theta))$ denote the Bregman divergence associated with this distribution. If*

1. *$d_\phi(y, \mu^0(\theta^0)) \le \epsilon$ (small noise),*

2. *$d_\phi(y, \mu^*(\theta^*)) \le \epsilon$ (constraint in the GLM problem in eq. 7.18), and*

3. *$\phi''(y)$ exists and is bounded on $[y_{min}, y_{max}]$, where $y_{min} = \min\{y, \mu^0, \mu^*\}$ and $y_{max} = \max\{y, \mu^0, \mu^*\}$,*

then

$$|\theta^* - \theta^0| \le \beta(\epsilon) = \sqrt{\epsilon} \cdot \frac{2\sqrt{2} \max_{\hat{\mu}\in[\mu^*;\mu^0]} |\phi''(\hat{\mu})|}{\sqrt{\min_{\hat{y}\in[y_{min};y_{max}]} \phi''(\hat{y})}}. \qquad (7.23)$$

Proof. We prove the lemma in two steps: first, we show that $|\mu^*(\theta^*) - \mu^0(\theta^0)|$ is small if ϵ is small, and then infer $|\theta^* - \theta^0|$ is small.

1. By definition in eq. 7.17, Bregman divergence is the non-linear tail of the Taylor expansion of $\phi(y)$ at point μ, i.e., the *Lagrange remainder* of the linear approximation:

$$d_\phi(y,\mu) = \phi''(\hat{y})(y-\mu)^2/2, \quad \hat{y} \in [y_1; y_2],$$

where $y_1 = \min\{y,\mu\}$, $y_2 = \max\{y,\mu\}$.

Let $y_1^0 = \min\{y,\mu^0\}$, $y_2^0 = \max\{y,\mu^0\}$, $y_1^* = \min\{y,\mu^*\}$, and $y_2^* = \max\{y,\mu^*\}$. Using the conditions $0 \le d_\phi(y,\mu^0) \le \epsilon$ and $0 \le d_\phi(y,\mu^*) \le \epsilon$, and observing that

$$\min_{\hat{y}\in[y_{min};y_{max}]} \phi''(\hat{y}) \le \min_{\hat{y}\in[y_1^0;y_2^0]} \phi''(\hat{y})$$

and

$$\min_{\hat{y}\in[y_{min};y_{max}]} \phi''(\hat{y}) \le \min_{\hat{y}\in[y_1^*;y_2^*]} \phi''(\hat{y}),$$

we get

$$\phi''(\hat{y})(y-\mu^0)^2/2 \le \epsilon \;\Leftrightarrow\; (y-\mu^0)^2 \le \frac{2\epsilon}{\phi''(\hat{y})} \;\Leftrightarrow$$

$$\Leftrightarrow |y-\mu^0| \le \frac{\sqrt{2\epsilon}}{\sqrt{\min_{\hat{y}\in[y_1^0;y_2^0]}\phi''(\hat{y})}} \le$$

$$\le \frac{\sqrt{2\epsilon}}{\sqrt{\min_{\hat{y}\in[y_{min};y_{max}]}\phi''(\hat{y})}}$$

and, similarly, $|y-\mu^*| \le \dfrac{\sqrt{2\epsilon}}{\sqrt{\min_{\hat{y}\in[y_1^*;y_2^*]}\phi''(\hat{y})}} \le$

$$\le \frac{\sqrt{2\epsilon}}{\sqrt{\min_{\hat{y}\in[y_{min};y_{max}]}\phi''(\hat{y})}},$$

from which, using the triangle inequality, we conclude

$$|\mu^*-\mu^0| \le |y-\mu^*| + |y-\mu^0| \le$$

$$\le \frac{2\sqrt{2\epsilon}}{\sqrt{\min_{\hat{y}\in[y_{min};y_{max}]}\phi''(\hat{y})}}. \tag{7.24}$$

Note that $\phi''(\hat{y})$ under the square root is always positive since ϕ is strictly convex.

2. The mean and the natural parameters of an exponential-family distribution relate to each other as follows: $\theta(\mu) = \phi'(\mu)$ (respectively, $\theta(\mu) = \nabla\phi(\mu)$ for vector μ), where $\phi'(\mu)$ is the link function. Therefore, we can write

$$|\theta^* - \theta^0| = |\phi'(\mu^*) - \phi'(\mu^0)| = |\phi''(\hat{\mu})(\mu^* - \mu^0)|,$$

where $\hat{\mu} \in [\mu^*; \mu^0]$,

and thus, using the above result in eq. 7.24, we get

$$|\theta^* - \theta^0| \leq \beta(\epsilon) = \sqrt{\epsilon} \cdot \frac{2\sqrt{2} \max_{\hat{\mu} \in [\mu^*; \mu^0]} |\phi''(\hat{\mu})|}{\sqrt{\min_{\hat{y} \in [y_{min}; y_{max}]} \phi''(\hat{y})}},$$

which concludes the proof.

$\square$

The condition (3) in the above lemma requires that $\phi''(y)$ exists and is bounded on the intervals between y and both μ^0 and μ^*. However, even when this condition is not satisfied, as it happens for the logistic loss, where $\phi''(y) = \frac{1}{y(1-y)}$ is unbounded at 0 and 1, and for several other Bregman divergences shown in Table 7.1, we may still be able to prove similar results using specific properties of each $\phi(y)$, as shown by the following lemmas.

Lemma 7.6. *(Bernoulli noise/Logistic loss) Let the conditions (1) and (2) of Lemma 7.5 be satisfied, and let $\phi(y) = y \log y + (1-y) \log(1-y)$, which corresponds to the logistic-loss Bregman divergence and Bernoulli distribution $p(y) = \mu^y (1 - \mu)^{1-y}$, where the mean parameter $\mu = P(y = 1)$. We assume that $0 < \mu^* < 1$, and $0 < \mu^0 < 1$. Then*

$$|\theta^0 - \theta^*| \leq \beta(\epsilon) = 4\epsilon.$$

Proof. Using the definition of the logistic-loss Bregman divergence from Table 1, and the conditions (1) and (2) of Lemma 7.5, we can write

$$d_\phi(y, \mu^0) = y \log(\frac{y}{\mu_0}) + (1 - y) \log(\frac{1-y}{1-\mu^0}) \leq \epsilon,$$

$$d_\phi(y, \mu^*) = y \log(\frac{y}{\mu_*}) + (1 - y) \log(\frac{1-y}{1-\mu^*}) \leq \epsilon, \qquad (7.25)$$

which implies

$$|d_\phi(y, \mu^0) - d_\phi(y, \mu^*)| \leq 2\epsilon, \qquad (7.26)$$

and, after substituting expressions 7.25 into eq. 7.26, and simplifying, we get

$$|y \log(\frac{\mu^0}{\mu^*}) + (1 - y) \log(\frac{1-\mu^0}{1-\mu^*})| \leq 2\epsilon. \qquad (7.27)$$

The above must be satisfied for each $y \in \{0, 1\}$ (the domain of Bernoulli distribution). Thus, we get

$$(1) \; |\log(\frac{1-\mu^0}{1-\mu^*})| \leq 2\epsilon \text{ if } y = 0, \text{ and}$$

$$(2) \; |\log(\frac{\mu^0}{\mu^*})| \leq 2\epsilon \text{ if } y = 1, \qquad (7.28)$$

or, equivalently

$$(1) \ e^{-2\epsilon} \leq \frac{1 - \mu^0}{1 - \mu^*} \leq e^{2\epsilon} \ \text{if} \ y = 0, \text{and}$$

$$(2) \ e^{-2\epsilon} \leq \frac{\mu^0}{\mu^*} \leq e^{2\epsilon} \ \text{if} \ y = 1.$$

Let us first consider the case of $y = 0$; subtracting 1 from the corresponding inequalities yields

$$e^{-2\epsilon} - 1 \leq \frac{\mu^* - \mu^0}{1 - \mu^*} \leq e^{2\epsilon} - 1 \Leftrightarrow$$

$$\Leftrightarrow (1 - \mu^*)(e^{-2\epsilon} - 1) \leq \mu^* - \mu^0 \leq (1 - \mu^*)(e^{2\epsilon} - 1).$$

By the mean value theorem, $e^x - 1 = e^x - e^0 = \frac{d(e^x)}{dx}|_{\hat{x}} \cdot (x - 0) = e^{\hat{x}}x$, for some $\hat{x} \in [0, x]$ if $x > 0$, or for some $\hat{x} \in [x, 0]$ if $x < 0$. Thus, $e^{-2\epsilon} - 1 = -e^{\hat{x}} \cdot 2\epsilon$, for some $\hat{x} \in [-2\epsilon, 0]$, and since e^x is a continuous monotone increasing function, $e^{\hat{x}} \leq 1$ and thus $e^{-2\epsilon} - 1 \geq -2\epsilon$. Similarly, $e^{2\epsilon} - 1 = e^{\hat{x}} \cdot 2\epsilon$, for some $\hat{x} \in [0, 2\epsilon]$, and since $e^{\hat{x}} \leq e^{2\epsilon}$, we get $e^{2\epsilon} - 1 \leq 2\epsilon \cdot e^{2\epsilon}$. Thus,

$$-2\epsilon(1 - \mu^*) \leq \mu^* - \mu^0 \leq 2\epsilon e^{2\epsilon}(1 - \mu^*) \Rightarrow$$

$$\Rightarrow |\mu^* - \mu^0| \leq 2\epsilon \cdot e^{2\epsilon}. \tag{7.29}$$

Similarly, in the case of $y = 1$, we get

$$e^{-2\epsilon} - 1 \leq \frac{\mu^0 - \mu^*}{\mu^*} \leq e^{2\epsilon} - 1,$$

and can apply the same derivation as above, and get the same result for $|\mu^* - \mu^0|$ as in eq. 7.29. Finally, since $\theta(\mu) = \phi'(\mu) = \log(\frac{\mu}{1-\mu})$, we get

$$|\theta^0 - \theta^*| = |\log(\frac{\mu^0}{1 - \mu^0}) - \log(\frac{\mu^*}{1 - \mu^*})| =$$

$$= |\log(\frac{\mu^0}{\mu^*}) - \log(\frac{1 - \mu^0}{1 - \mu^*})|.$$

From eq. 7.28 we get $|\log(\frac{\mu^0}{\mu^*})| \leq 2\epsilon$ and $|\log(\frac{1-\mu^0}{1-\mu^*})| \leq 2\epsilon$, which implies

$$|\theta^0 - \theta^*| = |\log(\frac{\mu^0}{\mu^*}) - \log(\frac{1 - \mu^0}{1 - \mu^*})| \leq 4\epsilon.$$

$\square$

Lemma 7.7. *(Exponential noise/Itakura-Saito distance) Let the conditions (1) and (2) of Lemma 7.5 be satisfied, and let $\phi(y) = -\log\mu - 1$, which corresponds to the Itakura-Saito distance $d_\phi(y, \mu) = \frac{y}{\mu} - \log(\frac{y}{\mu}) - 1$ and exponential distribution*

$p(y) = \lambda e^{\lambda y}$, *where the mean parameter* $\mu = 1/\lambda$. *We will also assume that the mean parameter is always separated from zero, i.e.* $\exists c_\mu > 0$ *such that* $\mu \geq c_\mu$. *Then*

$$|\theta^* - \theta^0| \leq \beta(\epsilon) = \frac{\sqrt{6}\,\epsilon}{c_\mu}.$$

Proof. To establish the result of the lemma we start with inequality $|u - \log u - 1| \leq \epsilon$, where u is $\frac{y}{\mu}$. Replacing u by $z = u - 1$, $z > -1$ gives us $|z - \log(1 + z)| \leq \epsilon$. Without loss of generality, let us assume that $\epsilon \leq \frac{1}{18}$. Then the Taylor decomposition of function $z - \log(1 + z)$ at the point $z = 0$,

$$z - \log(1 + z) = \frac{z^2}{2} - \frac{z^3}{3} + \frac{\theta^4}{4}, \text{ for } \theta \in [0, z] \text{ or } [z, 0],$$

implies that

$$\epsilon \geq z - \log(1 + z) \geq \frac{z^2}{2} - \frac{z^3}{3} \ (\text{ since } \frac{\theta^4}{4} \geq 0 \ ).$$

This, in turns, implies that $z \leq \frac{1}{3}$ and $\frac{z^2}{2} - \frac{z^3}{3} \geq \frac{z^2}{6}$ for $0 \leq z \leq \frac{1}{3}$.
    Hence

$$z - \log(1 + z) \geq \frac{z^2}{2} \quad \text{for} \quad -\frac{1}{3} \leq z \leq 0, \tag{7.30}$$

$$z - \log(1 + z) \geq \frac{z^2}{6} \quad \text{for} \quad 0 \leq z \leq \frac{1}{3}. \tag{7.31}$$

Combining together both estimates we get $|z| \leq \sqrt{6}\,\epsilon$, or

$$|y - \mu| \leq \sqrt{6}\,\epsilon \cdot \mu,$$

and

$$|\mu^0 - \mu^*| \leq \sqrt{6}\,\epsilon \cdot \max\{\mu^0, \mu^*\}.$$

Then

$$|\theta^* - \theta^0| = |\frac{1}{\mu^0} - \frac{1}{\mu^*}| = |\frac{\mu^* - \mu^0}{\mu^* \mu^0}| \leq \frac{\sqrt{6}\,\epsilon}{\min\{\mu^*, \mu^0\}} \leq \frac{\sqrt{6}\,\epsilon}{c_\mu},$$

since by the assumption of the lemma, $\min\{\mu^*, \mu^0\} \geq c_\mu$. $\qquad\square$

We now consider multivariate exponential-family distributions; the next lemma handles the general case of a multivariate Gaussian distribution (not necessarily the spherical one that had a diagonal covariance matrix and corresponded to the standard Euclidean distance (see Table 1).

Lemma 7.8. *(Non-i.i.d. Multivariate Gaussian noise/Mahalanobis distance) Let* $\phi(\mathbf{y}) = \mathbf{y}^T C \mathbf{y}$, *which corresponds to the general multivariate Gaussian with concentration matrix C, and Mahalanobis distance $d_\phi(y, \mu) = \frac{1}{2}(\mathbf{y} - \mu)^T C(\mathbf{y} - \mu)$. If $d_\phi(y, \mu^0) \leq \epsilon$ and $d_\phi(y, \mu^*) \leq \epsilon$, then*

$$||\theta^0 - \theta^*|| \leq \sqrt{2\epsilon}||C^{-1}||^{1/2} \cdot ||C||,$$

where $||C||$ is the operator norm.

Proof. Since C is (symmetric) positive definite, it can be written as $C = L^T L$, where L defines a linear operator on $\mathbf{y}$ space, and thus

$$\epsilon/2 \geq (\mathbf{y} - \mu)^T C(\mathbf{y} - \mu) = (L(\mathbf{y} - \mu))^T (L(\mathbf{y} - \mu)) =$$

$$= ||L(\mathbf{y} - \mu)||^2.$$

Also, it is easy to show that $||C^{-1}||I \leq C \leq ||C||I$ (where $||B||$ denotes the operator norm of B), and that

$$\epsilon/2 \geq ||L(\mathbf{y} - \mu)||^2 \geq ||L^{-1}||^{-2}||\mathbf{y} - \mu||^2 \Rightarrow$$

$$\Rightarrow ||\mathbf{y} - \mu|| \leq \sqrt{\frac{\epsilon}{2}}||L^{-1}||.$$

Then, using triangle inequality, we get

$$||\mu^* - \mu^0|| \leq ||\mathbf{y} - \mu^0|| + ||\mathbf{y} - \mu^*|| \leq \sqrt{2\epsilon}||L^{-1}||.$$

Finally, since $\theta(\mu) = \nabla\phi(\mu) = C\mu$, we get

$$||\theta^0 - \theta^*|| = ||C\mu^0 - C\mu^*|| \leq ||C|| \cdot ||\mu^0 - \mu^*|| =$$

$$= ||C|| \cdot ||\mu^0 - \mu^*|| \leq \sqrt{2\epsilon}||L^{-1}|| \cdot ||C||.$$

Note that $||L^{-1}|| = ||C^{-1}||^{1/2}$, which concludes the proof. $\square$

7.4 Summary and Bibliographic Notes

In this section, we discussed an extension of standard results on noisy sparse signal recovery (Candès et al., 2006b) to the general case of *exponential-family noise*, where the LASSO problem is replaced by the l_1-regularized *Generalized Linear Model (GLM)* regression. As shown in (Rish and Grabarnik, 2009), under standard restricted isometry property (RIP) assumptions on the design matrix, l_1-norm minimization can provide a stable recovery of a sparse signal under exponential-family noise assumptions, provided that the noise is sufficiently small and the distribution satisfies certain (sufficient) conditions, such as bounded second derivative of the Legendre conjugate $\phi(y)$ of the log-partition function that uniquely determines the distribution. Also, distribution-specific proofs are provided for several members of the exponential family that may not satisfy the above general condition.

Theoretical analysis of regularized maximum-likelihood estimators, or M-estimators, became a very popular topic in recent years. A large number of novel theoretical results have been derived, addressing different aspects of their behavior, including the model-selection consistency, i.e. an accurate recovery of nonzero pattern of predictor variables, l_2- and l_1-norm error in parameter estimation (as discussed in this chapter), or prediction error.

For example, recent work on l_1-norm regularized Generalized Linear Models (GLMs) includes results on risk (expected loss) consistency (van de Geer, 2008), model-selection consistency of logistic regression (Bunea, 2008; Ravikumar et al., 2010), and consistency in l_2- and l_1-norm (Bach, 2010; Kakade et al., 2010). More specifically, using the properties of self-concordant functions (Boyd and Vandenberghe, 2004) and following the technique of Restricted Eigenvalues developed in (Bickel et al., 2009), a multidimensional version of the result similar to 7.6 was established in Theorem 5 of (Bach, 2010). Also, (Kakade et al., 2010) use the technique of Restricted Eigenvalues and (almost) strong convexity of the Fisher risk for the sub-Gaussian random noise to obtain the result (Corollary 4.4) similar to 7.6.

Recent work by (Negahban et al., 2009), and its extended version presented in (Negahban et al., 2012), considers a unifying framework for analysis of regularized maximum-likelihood estimators (M-estimators), and states sufficient conditions that guarantee asymptotic recovery (i.e. consistency) of sparse models' parameters (i.e., sparse signals). These general conditions are: *decomposability* of the regularizer (which is satisfied for l_1-norm), and *restricted strong convexity (RSC)* of the loss function. Generalized Linear Models are considered as a special case, and consistency results from GLMs are derived from the main result using the above two sufficient conditions. Since the l_1-regularizer is decomposable, the main challenge is establishing RSC for the exponential-family negative log-likelihood loss, and this is achieved by imposing two (sufficient) conditions, called GLM1 and GLM2, on the design matrix and on the exponential-family distribution, respectively. Briefly, the GLM1 condition requires the rows of the design matrix to be i.i.d. samples with sub-Gaussian behavior, and the GLM2 condition includes as one of the alternative sufficient conditions a uniformly bounded second derivative of the cumulant function, similar to our Lemma 7.5 (which bounds its Legendre conjugate). Given GLM1 and GLM2 conditions, (Negahban et al., 2012) derive a bound on l_2-norm of the difference between the true signal and the solution to l_1-regularized GLM regression. The result is probabilistic, with the probability approaching 1 as the number of samples increases. Note, however, some differences between those results and the earlier results of (Rish and Grabarnik, 2009) summarized in this chapter. First, the bounds presented here are deterministic and the design matrix must satisfy RIP rather than sub-Gaussianity. Second, the focus here is on the constrained l_1-norm minimization formulation rather than on its Lagrangian form; in the constrained formulation, parameter ϵ bounding the divergence between the linear projections of the signal and its noisy observations (e.g., $||\mathbf{y} - A\mathbf{x}||_{l_2} < \epsilon$) has a clear intuitive meaning, characterizing the amount of noise in measurements, while the particular values of the sparsity parameter λ in Lagrangian formulation are somewhat harder to interpret. The results summarized here provide a very intuitive and straightforward extension of the standard compressed sensing result presented in (Candès et al., 2006b).

Regarding the algorithms available for solving the l_1-regularized Generalized Linear Model (GLM) regression, one of the popular techniques is the LARS-like path-following method for GLMs introduced by (Park and Hastie, 2007). It uses the predictor-corrector method of convex optimization, which finds a series of solutions for a varying parameter of the optimization problem (in this case, the sparsity parameter in front of the l_1-norm) by using the initial conditions (solutions at one extreme value of the parameter) and continuing to find the adjacent solutions on the basis of the current solutions. The method of (Park and Hastie, 2007) generalizes the LARS idea to the GLM path which, unlike the LARS/Lasso path, is not piecewise linear.

Another large body of recent research is devoted to regularized M-estimators that arise in learning probabilistic graphical models, such as Markov Networks, or Markov Random Fields, discussed in the next chapter.

Chapter 8

Sparse Graphical Models

In many practical applications such as social network analysis, reverse-engineering of gene networks, or discovering functional brain connectivity patterns, just to name a few, the ultimate objective is to reconstruct underlying dependencies among the variables of interest, such as individuals, genes, or brain areas. Probabilistic graphical models provide a convenient visualization and inference tool that captures statistical dependencies among random variables explicitly in a form of a graph.

A common approach to learning probabilistic graphical models is to choose the simplest model, e.g., the sparsest network, that adequately explains the data. Formally, this leads to a regularized maximum-likelihood problem with the penalty on the number of parameters (i.e., the l_0-norm), which is an intractable combinatorial problem, in general. However, similarly to the regression setting considered in the previous chapters, replacing the intractable l_0-norm by its convex relaxation via the l_1-norm is a popular approach to tackling the sparse graphical model learning problem. Note that, as discussed before, the sparsity requirement not only improves the interpretability of the model, but also serves as a regularizer that helps to avoid overfitting in cases of high-dimensional problems with a limited number of samples.

In this chapter, we focus on *undirected* graphical models, known as *Markov networks*, or *Markov random fields (MRFs)*; more specifically, we consider Gaussian MRFs, i.e. MRFs defined over Gaussian random variables. In the past decade, learning sparse Gaussian MRFs (GMRFs) became a rapidly growing area, and multiple efficient optimization approaches were proposed for learning such models. We will review some of those algorithms and discuss applications of sparse GMRF learning in neuroimaging. We will also discuss an important practical issue of regularization parameter selection, i.e. choosing the "right" level of sparsity. Clearly, sparse graphical model learning is a very large field, and there are multiple other recent developments, such as, for example, sparse discrete-variable (e.g., binary) Markov networks, as well as sparse directed graphical models, or Bayesian networks, that remain out of the scope of this chapter. The bibliography section provides several references to some of the recent advances in those fields.

We will now start with a review of some basic concepts related to probabilistic graphical models.

8.1 Background

A *graph* G is a pair $G = (V, E)$, where V is a finite set of *vertices*, or *nodes*, and $E \subseteq V \times V$ a set of *edges*, or *links*, that connect pairs of nodes $(i, j) \in E$. An edge between the nodes $i \in V$ and $j \in V$ is *undirected* if $(i, j) \in E$ and $(j, i) \in E$, otherwise it is called a *directed* edge. G is an *undirected graph* if all its edges are undirected, and G is a *directed graph* if all its edges are directed.

A *loop*, or a *self-loop*, is an edge that connects a node to itself. A *simple graph* is a graph containing no self-loops and no *multiple edges* (the latter is true by the definition of E as a set, since a set contains only a single instance of each of its elements). From now on, when we refer to a *graph*, we will always assume a simple graph.

Two nodes i and j are *adjacent* to each other if they are connected by an edge $(i, j) \in E$. They are also called *neighbors* of each other; the set of all neighbors $Ne(i)$ of a node i is called its *neighborhood* in G. A *path* from node i_1 to a node i_k is a sequence of distinct nodes $i_1, i_2,...,i_k$, such that each pair of subsequent nodes i_j and i_{j+1} is connected by an edge $(i, j) \in E$.

A *complete* graph is a graph where each pair of nodes is connected by an edge. A *subgraph* G_S of G is a graph over a subset of nodes $S \subseteq V$ and the subset of edges $E_S \subseteq E$ that connects nodes in S, i.e., $(i, j) \in E_S$ if and only if $i \in S$, $j \in S$, and $(i, j) \in E$. The subgraph G_S is said to be *induced* by the nodes in S. A *clique* C is a complete subgraph of G. A *maximal clique* is a clique that is not a subgraph of any larger clique.

Let $\mathbf{X}$ be a set of *random variables* associated with a *probability measure* $p(\mathbf{X})$. The probability measure is also called the *(joint) probability distribution*, in the case of discrete variables, or the *(joint) probability density*, in the case of continuous variables. A vector $\mathbf{x}$ will denote a particular *value assignment* to the variable in $\mathbf{X}$, also called a *state*, or a *configuration*.

Given three *disjoint* subsets of random variables, $\mathbf{X}_1 \subset \mathbf{X}$, $\mathbf{X}_2 \subset \mathbf{X}$, and $\mathbf{X}_3 \subset \mathbf{X}$, the subsets $\mathbf{X}_1$ and $\mathbf{X}_2$ are said to be *conditionally independent* given $\mathbf{X}_3$ (denoted $\mathbf{X}_1 \perp\!\!\!\perp \mathbf{X}_2 | \mathbf{X}_3$) if and only if $p(\mathbf{X}_1, \mathbf{X}_2 | \mathbf{X}_3) = p(\mathbf{X}_1 | \mathbf{X}_3)p(\mathbf{X}_2 | \mathbf{X}_3)$, or, equivalently, $p(\mathbf{X}_1 | \mathbf{X}_2, \mathbf{X}_3) = p(\mathbf{X}_1 | \mathbf{X}_3)$. When $\mathbf{X}_3 = \emptyset$, $\mathbf{X}_1$ and $\mathbf{X}_2$ are said to be *marginally independent*.

A *probabilistic graphical model* is a triplet $(\mathbf{X}, p(\mathbf{X}), G)$, where $\mathbf{X}$ is a set of random variables, associated with a probability measure $p(\mathbf{X})$, and $G = (V, E)$ is a graph, where the nodes in V are in a one-to-one correspondence with the variables in $\mathbf{X}$, and the edges in E are used to encode probabilistic dependence and independence relations among the variables. Their precise meaning will be defined below, in the specific context of directed and undirected graphical models.

Graphical models have three main advantages: they (1) provide a convenient tool for visualizing statistical dependencies and independencies among sets of random variables, (2) encode joint probabilities over large sets of variables in a compact way, using factorized representations according to the graph structure, and (3) allow for efficient graph-based inference algorithms. There are several types of probabilistic graphical models, such as undirected models (*Markov networks*), directed models (*Bayesian networks*), and a general class of *chain graphs* containing both directed and undirected edges. These models have different advantages and drawbacks, in terms of their "expressive power", i.e. probabilistic independence relationships that they are capable of representing, convenience of factorization, and inference algorithms; for example, certain sets of independence assumptions are better expressed by a Markov network rather than a Bayesian network, and vice versa. For a comprehensive treatment of graphical models see, for example, (Lauritzen, 1996; Pearl, 1988; Koller and Friedman, 2009).

8.2 Markov Networks

We now introduce undirected graphical models called Markov networks, starting with the most commonly used class of those models that *factorize* according to the graph cliques[1]:

Definition 10. *A **Markov network**, also called a **Markov random field (MRF)**, is an **undirected** graphical model $(\mathbf{X}, p(\mathbf{X}), G)$ representing a joint probability*

---

[1]Note that a more general definition of Markov networks stated in terms of the network's Markov properties is discussed in section 8.2.1 below (including relatively rare cases of distributions that do not factorize). However, establishing these properties for an arbitrary probability distribution is not always easy and/or practical. Also, as shown in the next section, the factorization property stated here implies these Markov properties; furthermore, factorization property is equivalent to Markov properties (see Hammersley-Clifford theorem, section 8.2.1) when the probability density is strictly positive (i.e., there are no zero-probability states) – which covers most of the typical real-life applications.

distribution that factorizes as follows:

$$p(\mathbf{x}) = \frac{1}{Z} \prod_{C \in Cliques} \phi_C(\mathbf{X}_C), \qquad (8.1)$$

where Cliques is the set of all maximal cliques in G, $\phi_C(\mathbf{X}_C)$ are nonnegative **potential functions** *defined over the subsets of the variables corresponding to each clique, and $Z = \sum_{\mathbf{x}} \prod_{C \in Cliques} \phi_C(\mathbf{X}_C)$ is the normalization constant, also called the* **partition function**.

Given a distribution factorized according to the graph G as stated in eq. 8.1, the graph satisfies the following three *Markov properties*: *pairwise Markov property* (any two non-adjacent variables are conditionally independent given all other variables), *local Markov property* (a variable is conditionally independent of all other variables given its neighbors), and *global Markov property* (any two subsets of variables are conditionally independent given a separating subset); these Markov properties are formally defined in section 8.2.1 below.

8.2.1　Markov Network Properties: A Closer Look

We will now present a more general definition of a Markov network stated in terms of the graph properties, and discuss the relationship between Markov properties and the factorization property introduced in the previous section. Note that the material of this section, provided for the completeness sake, is not necessary for understanding the remainder of this chapter, and thus can be skipped if needed.

Most generally, a *Markov network* is defined as an *undirected* graphical model $(\mathbf{X}, p(\mathbf{X}), G)$, where the undirected graph G satisfies the *global Markov property* (called (G) property), stating that any two subsets of variables $\mathbf{Y} \subset \mathbf{X}$ and $\mathbf{Z} \subset \mathbf{X}$ are independent given the third subset $\mathbf{S} \subset \mathbf{X}$, if $\mathbf{S}$ *separates* $\mathbf{Y}$ and $\mathbf{Z}$ in G, i.e., any path from a node in $\mathbf{Y}$ to a node in $\mathbf{Z}$ contains a node from $\mathbf{S}$:

$$(G) \quad \forall \mathbf{S}, \mathbf{Y}, \mathbf{Z} \subset V, \ \mathbf{S} \ separates \ \mathbf{Y} \ and \ \mathbf{Z} \Rightarrow X_{\mathbf{Y}} \perp\!\!\!\perp X_{\mathbf{Z}} | X_{\mathbf{S}}, \qquad (8.2)$$

where $\mathbf{X}_A \subseteq \mathbf{X}$ denotes a subset of variables corresponding to a subset of nodes $\mathbf{A} \subseteq \mathbf{V}$.

An undirected graph satisfying the property (G) is also called an *independence map*, or *I-map*, of the distribution $p(\mathbf{X})$. Obviously, a complete graph is a trivial I-map for any distribution. Thus, a Markov network is often defined as a *minimal I-map* (see, e.g., (Pearl, 1988)), i.e. no edge can be removed from it without violating its I-map property. Note, however, that, given a (minimal) I-map, the reverse of the above condition in eq. 8.2:

$$\text{D-map}: \quad \forall \mathbf{S}, \mathbf{Y}, \mathbf{Z} X_{\mathbf{Y}} \perp\!\!\!\perp X_{\mathbf{Z}} | X_{\mathbf{S}} \Rightarrow \subset V, \ \mathbf{S} \ separates \ \mathbf{Y} \ and \ \mathbf{Z}, \qquad (8.3)$$

does not necessarily hold, i.e. there are some sets of independence relations implied by a probability distribution that may not be represented by an I-map. In case when condition 8.3 holds, the graph is called a *dependency map*, or *D-map*. A graph is called called a *perfect map*, or *P-map* of a distribution $P(X)$, if it is both an I-map and D-map of that distribution.

There are also two other commonly used Markov properties that relate graph constraints and statistical independencies. The *local Markov property* (called (L) property) implies that a variable X_i is conditionally independent of the rest of the variables given its neighbors $Ne(i)$ in G:

$$(L): \quad X_i \perp\!\!\!\perp X_{V/\{i \cup Ne(i)\}} | X_{Ne(i)}, \tag{8.4}$$

while the *pairwise Markov property* (called (P) property) states that the lack of edge (i, j) implies conditional independence between X_i and X_j given the remaining variables (Lauritzen, 1996):

$$(P): \quad \{i, j\} \notin E \Rightarrow X_i \perp\!\!\!\perp X_j | X_{V/\{i,j\}}. \tag{8.5}$$

In general, the three Markov properties defined above do not necessarily coincide (see (Lauritzen, 1996) for examples); moreover, it is easy to show that

$$(G) \Rightarrow (L) \Rightarrow (P). \tag{8.6}$$

However, *all three Markov properties become equivalent if the probability measure is strictly positive* (Pearl, 1988; Lauritzen, 1996), i.e., if there are no zero-probability states ($p(\mathbf{x}) > 0$ for all $\mathbf{x}$). This equivalence follows from the result by (Pearl and Paz, 1987), and is discussed in various textbooks (Lauritzen, 1996; Pearl, 1988; Cowell et al., 1999).

The strict positivity condition is frequently satisfied in practice, and is also important for establishing a link between the *qualitative* part of a probability distribution, i.e. its graphical structure (independence relations), and its *quantitative* part, i.e. parameters (state probabilities). In order to make Markov networks useful, we need a compact way of specifying the state probabilities over large number of variables. Factorization according to the graph structure provides such compact representation. We say that the probability measure satisfies the *factorization property*, or (F) property, with respect to a graph G, if

$$p(\mathbf{x}) = \frac{1}{Z} \prod_{C \in Cliques} \phi_C(\mathbf{X}_C), \tag{8.7}$$

where $Cliques$ is the set of all maximal cliques in G, $\phi_C(\mathbf{X}_C)$ are the nonnegative *potential functions* defined over the subsets of the variables corresponding to each clique, and $Z = \sum_{\mathbf{x}} \prod_{C \in Cliques} \phi_C(\mathbf{X}_C)$ is the normalization constant, also called the *partition function*. A commonly considered special case of Markov networks is the *pairwise Markov network*, where the potential functions are defined over the pairs of nodes connected by an edge. Thus, in a pairwise Markov network, a potential function over the clique L, B, and D would factorize as $\phi(L, B, D) = \phi_1(L, B)\phi_1(B, D)\phi_1(L, D)$.

In general, the factorization property implies the Markov properties.

Theorem 8.1. *(Lauritzen, 1996) For any undirected graph G and any probability measure on $p(\mathbf{X})$,*

$$(F) \Rightarrow (G) \Rightarrow (L) \Rightarrow (P). \tag{8.8}$$

However, the factorization property and the Markov properties become equivalent in cases of strictly positive $p(\mathbf{X})$. A strictly positive probability measure that factorizes according to a graph G as specified in eq. 8.7 is also called a *Gibbs random field (GRF)* (the name has roots in statistical physics). The following key result establishes the equivalence between MRFs and GRFs for strictly positive $p(\mathbf{X})$:

Theorem 8.2. *(Hammersley and Clifford, 1971) A strictly positive probability measure $p(\mathbf{X})$ factorizes over a graph G if and only if it satisfies Markov properties with respect to G:*

$$(F) \Leftrightarrow (G) \Leftrightarrow (L) \Leftrightarrow (P), \tag{8.9}$$

i.e., $p(\mathbf{X}) > 0$ is an MRF if and only if it is a GRF.

This result was first proved in (Hammersley and Clifford, 1971) for discrete-valued random variables. More recent and simpler proofs are also given in (Grimmett, 1973; Preston, 1973; Sherman, 1973; Besag, 1974; Moussouris, 1974; Kindermann and Snell, 1980; Lauritzen, 1996). The idea is to use the Möbius inversion in order to show $(P) \Rightarrow (F)$. Note that the strict positivity condition is essential: (Moussouris, 1974) provides an example of a four-node Markov network (a "square", i.e. a four-cycle without a chord) where the global Markov property (G) *does not* imply the factorization property when positivity assumption is violated.

Finally, we would like to mention that the definition of a Markov network can vary from one publication to another: some include strict positivity as a necessary part of the definition; some define a Markov network as a nonnegative distribution that factorizes, and then relate it to Markov properties under positivity conditions; some use the property (P) as a definition of a Markov network. Since the positivity assumptions is practically always made in the Markov network literature, it does not matter which definition was used in the beginning. Also, a Markov network is often defined as an I-map (i.e., satisfying the global Markov property), but the minimality condition (minimal I-map) is not always mentioned explicitly. In this book, we defined a Markov network as a minimal I-map, following the definition of (Pearl, 1988).

8.2.2   Gaussian MRFs

An important particular type of MRFs is a *Gaussian MRF (GMRF)*, i.e. an MRF encoding multivariate Gaussian distributions. Multivariate Gaussian distributions are often used for modeling continuous variables in practical applications, due to their well-understood mathematical properties that lead to simpler theoretical analysis and to computationally efficient learning algorithms.

Recall that a multivariate Gaussian density over a set of random variables $\mathbf{X}$ is defined as

$$p(\mathbf{x}) = (2\pi)^{-p/2} \det(\Sigma)^{-\frac{1}{2}} e^{-\frac{1}{2}(\mathbf{x}-\mu)^T \Sigma^{-1}(\mathbf{x}-\mu)}, \tag{8.10}$$

where μ is the *mean* and Σ is the *covariance matrix* of the distribution, respectively. Since $det(\Sigma)^{-1} = det(\Sigma^{-1})$, we can rewrite 8.10, denoting by $C = \Sigma^{-1}$ the inverse covariance matrix, also known as the *precision matrix*, or the *concentration matrix*:

$$p(\mathbf{x}) = (2\pi)^{-p/2} \det(C)^{\frac{1}{2}} e^{-\frac{1}{2}(\mathbf{x}-\mu)^T C(\mathbf{x}-\mu)}. \tag{8.11}$$

As it is shown in (Lauritzen, 1996), *a pair of Gaussian variables X_i and X_j is conditionally independent given the rest of the variables if and only if the corresponding entries in C are zero*, i.e. if and only if $c_{ij} = c_{ji} = 0$. Thus, missing edges in a Gaussian MRF imply zero entries in C. The reverse is also true, i.e. $c_{ij} = 0$ implies missing edges between X_i and X_j in the Gaussian MRF, although this may not be true for MRFs in general. Therefore, *learning the structure of a Gaussian MRF model is equivalent to identifying zero entries in the corresponding concentration matrix*, and will be the primary focus of section 8.4.

8.3 Learning and Inference in Markov Networks

Two main question arise in the context of probabilistic graphical models:

- How do we construct such models?

- How do we use such models?

The first question typically relates to *learning*, or *estimation*, of a graphical model from data, especially when the number of variables is large, e.g., on the order of hundreds or thousands, which is typical in modern applications such as biological or social networks. In such applications, constructing a graphical model manually, using only a domain expert knowledge, may not feasible, especially when the domain knowledge is limited and the goal is to actually gain novel insights from the data.

The second question relates to making *probabilistic inferences* based on a graphical model. While the main focus of this chapter is on learning graphical models, we will briefly discuss some inference problems later in this section, illustrating successful use of Markov networks in practical applications.

8.3.1 Learning

The problem of learning probabilistic graphical models has a long and rich history in statistics and machine learning (see the Bibliographic References at the end of this chapter). This problem involves learning the *structure* of a graphical model, i.e. *model selection*, and learning the *parameters* of the joint probability distribution, i.e. *parameter estimation*, which can be performed either after or during the structure learning step, depending on the method used. A commonly used approach to graphical model learning is *regularized likelihood maximization*, where the regularization penalizes the model complexity in some way. Traditional model selection criteria, such as AIC, BIC/MDL, and similar ones, use the penalty proportional to the number of model parameters, i.e. the l_0-norm of the parameter vector. Since finding the simplest (minimal l_0-norm) model that fits the data well is an NP-hard problem, approximate approaches such as greedy optimization were typically used in the past (Heckerman, 1995).

An alternative approach that became extremely popular in the past few years is to use instead a tractable relaxation of the above optimization problem. Similarly to sparse regression, the intractable l_0-norm optimization is replaced by its convex l_1-relaxation (Meinshausen and Bühlmann, 2006; Wainwright et al., 2007; Yuan and Lin, 2007; Banerjee et al., 2008; Friedman et al., 2007b). We will discuss these types of approaches to both continuous and discrete Markov networks later in this chapter.

8.3.2 Inference

Once a probabilistic graphical model is constructed, it can be used to make probabilistic inferences about the variables of interest, such as finding the probability of some unobserved variable(s) given the observed ones (e.g., is a certain gene likely to be expressed given the expressions of some other genes), or predicting the most likely state in classification problems (e.g., discriminating between healthy and sick subjects from brain imaging data).

Formally, let $\mathbf{Z} \subset \mathbf{X}$ be a subset of *observed* random variables that are assigned the values $\mathbf{Z} = \mathbf{z}$, and let $\mathbf{Y} \subseteq \mathbf{X} - \mathbf{Z}$ be a set of unobserved variables of interest. Then the probabilistic inference task is to find the posterior probability $P(\mathbf{Y}|\mathbf{Z} = \mathbf{z})$.

Probabilistic inference can be used for predicting the unobserved *response* variable (or *class label*) Y given the set of observed *features* $\mathbf{Z}$, in a machine-learning setting. Given a *training dataset* consisting of *training samples*, i.e. assignments to both the features and the response variable, we can learn a probabilistic graphical model; then, given a *test sample* consisting of a feature assignment, we use probabilistic inference to predict an unobserved response, or class label.

For example, in a particular case of the *classification* problem, $Y \in \mathbf{X}$ is a discrete variable (often binary). The probabilistic classification task is a decision problem, which involves computing $P(Y|\mathbf{Z} = \mathbf{z})$ and selecting the most likely class label $y^* = \arg\max_y P(Y = y|\mathbf{Z} = \mathbf{z})$. Using the *Bayes rule*, we get

$$P(Y = y|\mathbf{Z} = \mathbf{z}) = \frac{P(\mathbf{Z} = \mathbf{z}|Y = y)P(Y = y)}{P(\mathbf{Z} = \mathbf{z})} \qquad (8.12)$$

for each assignment $Y = y$. Since the denominator does not depend on Y, we only need to compute

$$y^* = \arg\max_y P(\mathbf{Z} = \mathbf{z}|Y = y)P(Y = y). \qquad (8.13)$$

Thus, given the training data, we can learn the models $P(\mathbf{Z} = \mathbf{z}|Y = y)P(Y = y)$, e.g., Markov networks over the set of features $\mathbf{Z}$, separately for each class label $Y = y$, and then, given an unlabeled test sample, assign the most likely class using eq. 8.13.

8.3.3 Example: Neuroimaging Applications

Let us consider an example of using Markov networks for predicting mental states of a person based on brain imaging data, such as functional MRI (fMRI). The dataset, first presented in (Mitchell et al., 2004), consists of a series of trials in which the subject was shown either a picture or a sentence. A subset of 1700 to 2200 voxels, dependent on a particular subject, was extracted based on the prior knowledge about the brain areas potentially relevant to the task. The voxels correspond to the features: each feature is the voxel signal averaged over 6 scans taken while a subject is presented with a particular stimulus, such as a picture or a sentence. The data contain 40 samples, where half of the samples correspond to the picture stimulus (+1) and the remaining half correspond to the sentence stimulus (-1).

A successful application of a sparse Markov network classifier to this dataset was described in (Scheinberg and Rish, 2010). For each class $Y = \{-1, 1\}$, a sparse Gaussian Markov network model was estimated from data using (see the next section for details); this provided an estimate of the Gaussian conditional density $p(\mathbf{x}|y)$, where $\mathbf{x}$ is the feature (voxel) vector; on the test data, the most-likely class label $\arg\max_y p(\mathbf{x}|y)P(y)$ was selected for each unlabeled test sample. Figure 8.1 presents the classification results for three different subjects, and for the increasing number of top-ranked voxels included in the model, where the ranking used the p-value in a two-sample t-test performed for each voxel (i.e., lower p-value reflects higher discriminative ability of a voxel and thus corresponds to a higher rank). Each point on the graph represents an average classification error over 40 samples when using leave-one-out cross-validation. A Markov network classifier is based on sparse Gaussian MRF models learned by an algorithm SINCO described later in

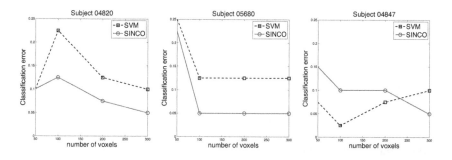

FIGURE 8.1: Using a sparse Gaussian Markov network classifier to predict cognitive states of a subject from fMRI data, such as reading a sentence versus viewing a picture.

this chapter. We see that the Markov network classifier yields quite accurate predictions for this task, achieving just 5% misclassification error, and is comparable to, and frequently better than, the state-of-the-art support-vector machine (SVM) classifier.

Another successful application of Markov network classifiers to mental state prediction was described in (Cecchi et al., 2009; Rish et al., 2013). The objective of that work was to discover predictive features (*statistical biomarkers*) and build a predictive statistical model of schizophrenia, a complex psychiatric disorder that has eluded a characterization in terms of local abnormalities of brain activity, and is hypothesized to affect the collective, "emergent" working of the brain. The dataset consisted of fMRI scans collected for schizophrenic and healthy subjects performing a simple auditory task in the scanner. As it was demonstrated in (Cecchi et al., 2009; Rish et al., 2013), topological features of a brain's *functional networks*, obtained from thresholded correlation matrices over the whole brain, contain a significant amount of information, allowing for accurate discrimination between the two groups of subjects. A *functional network* is a graph where the nodes correspond to voxels, and a link exists between a pair of nodes if their BOLD (blood-oxygenation-level dependent) signals are highly correlated (either positively or negatively), i.e., the absolute value of their correlation exceeds a given threshold ϵ. (In other words, the adjacency matrix of a functional network is simply the correlation matrix over the set of voxels, where the entries with absolute values exceeding ϵ are replaced by 1s and the others are replaced by 0s. Note that the functional network can be very different from the corresponding Markov network, since the latter reflects the sparsity pattern of the *inverse* covariance matrix.) In (Cecchi et al., 2009; Rish et al., 2013), voxel degrees in the functional network were shown to be highly predictive features, allowing for 86% accurate classification between schizophrenic and healthy subjects, when using leave-one-subject-out cross-validation on 44 samples, 2 samples (experiment runs) per subject, where half of the subjects where schizophrenic patients. Figure 8.2b demonstrates that the Gaussian Markov network (MRF) classifier was capable of achieving this accuracy only with a few hundreds of most-discriminative voxels, where two-sample t-test was used to rank voxels by their discriminative ability; Figure 8.2a shows the location of the most-discriminative voxels (about 1000 of them) that survived the False-Discovery Rate (FDR) correction for multiple comparison, a necessary step to rule out spurious results when performing hypothesis testing on more than 50,000 voxels. Gaussian MRF classifier significantly outperformed Gaussian Naive Bayes and linear SVM, apparently capturing interactions among the variables (voxel degrees) that turned out to be discriminative between the two groups.

Sparse Markov network models were also applied recently to the analysis of several other brain disorders, such as Alzheimer's disease (Huang et al., 2009) and drug addiction (Honorio et al., 2009, 2012). These studies allowed to identify certain abnormal changes in the network structures associated with the corresponding diseases. For example, (Honorio et al., 2012) further improves upon the standard sparse Markov network formulation by allowing node selection, besides the edge selection. This is achieved via sparse group-Lasso type of regularization similar to

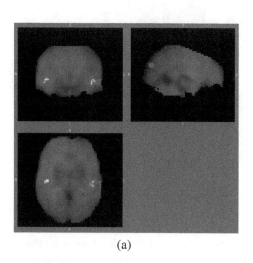

MRF vs GNB vs SVM:
schizophrenic vs normal

- MRF (0.1): degree (long–distance)
- GNB: degree (long–distance)
- SVM:degree (long–distance)

K top voxels (ttest)

(a)   (b)

FIGURE 8.2 (See color insert): (a) FDR-corrected 2-sample t-test results for (normalized) degree maps, where the null hypothesis at each voxel assumes no difference between the schizophrenic vs. normal groups. Red/yellow denotes the areas of low p-values passing FDR correction at $\alpha = 0.05$ level (i.e., 5% false-positive rate). Note that the mean (normalized) degree at those voxels was always (significantly) *higher* for normals than for schizophrenics. (b) Gaussian MRF classifier predicts schizophrenia with 86% accuracy using just 100 top-ranked (most-discriminative) features, such as voxel degrees in a functional network.

(Friedman et al., 2010) (i.e., combining group-Lasso penalty with the basic l_1-penalty); the groups here correspond to sets of edges adjacent to the same variable, thus setting all of them to zero as a group amounts to elimination of the node from the graph. This can improve the interpretability of the Markov network dramaticaly, especially in high-dimensional datasets where the number of nodes is on the order of thousands. For example, Figure 8.3, reproduced here from (Honorio et al., 2012), demonstrates the advantages of the sparse-group (variable-selection) approach to learning of sparse Markov networks versus the standard (edge-sparse) formulation outlined in the next section. Figure 8.3 shows the network structures learned for cocaine-addicted vs healthy control subjects, comparing the two methods. The disconnected variables are not shown. The variable-selection sparse Markov network approach yields much fewer connected variables but a higher log-likelihood than graphical lasso, as reported in (Honorio et al., 2012), which suggests that the discarded edges from the disconnected nodes are not important for accurate modeling of this dataset. Moreover, removal of a large number of nuisance variables (voxels)

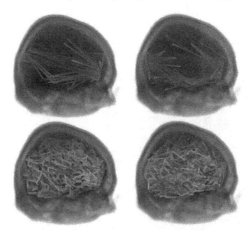

FIGURE 8.3 (See color insert): Structures learned for cocaine addicted
(left) and control subjects (right), for sparse Markov network learning method with variable-selection via $\ell_{1,2}$ method (top), and without
variable-selection, i.e., standard graphical lasso approach (bottom). Positive
interactions are shown in blue, negative interactions are shown in red. Notice that
structures on top are much sparser (density 0.0016) than the ones on the bottom
(density 0.023) where the number of edges in a complete graph is $\approx$378,000.

results into a much more interpretable model, clearly demonstrating brain areas involved in structural model differences that discriminate cocaine addicts from healthy
control subjects. Note that, as shown at the bottom of Figure 8.3, the standard approach to sparse Markov network learning, such as the glasso algorithm described
later in section 8.4.2, connects most of the brain voxels in both populations, making
it practically impossible to detect any network differences between the two groups
of subjects. The group-penalty approach produces much more "localized" networks
(top of Figure 8.3) that involve a relatively small number of brain areas: cocaine addicts show increased interactions between the visual cortex (back of the brain, on the
left here) and the prefrontal cortex (front of the brain image, on the right), while at
the same time decreased density of interactions between the visual cortex and other
brain areas (more clearly present in healthy control subjects). The alteration in this
pathway in the addict group is highly significant from a neuroscientific perspective.
First, the trigger for reward was a visual stimulus. Abnormalities in the visual cortex
were reported in Lee et al. (2003) when comparing cocaine abusers to control subjects. Second, the prefrontal cortex is involved in higher-order cognitive functions
such as decision making and reward processing. Abnormal monetary processing in
the prefrontal cortex was reported in Goldstein et al. (2009) when comparing cocaine addicted individuals to controls. Although a more careful interpretation of the
observed results remains to be done in the near future, these results are encouraging
and lend themselves to specific neuroscientific hypothesis testing.

8.4 Learning Sparse Gaussian MRFs

We will now discuss in detail the problem of learning sparse Gaussian MRFs, an important subclass of MRFs that is widely used in practice and lends itself to efficient optimization approaches. Recall that learning the structure of a Gaussian MRF model is equivalent to identifying zero entries in the corresponding precision, or inverse covariance, matrix (see section 8.2.2). This problem was first introduced by (Dempster, 1972), and is often referred to as the *covariance selection problem* (Dempster, 1972) or the *model-selection problem in the Gaussian concentration graph model* (Cox and Wermuth, 1996).

We assume the dataset consisting of n independent and identically distributed (i.i.d.) samples $D = \{\mathbf{x}_1,...,\mathbf{x}_n\}$, where each sample is a p-dimensional vector – an assignment to the variables in $\mathbf{X}$. The log-likelihood of such dataset can be written as:

$$L(D) = \frac{n}{2}\det(C) - \frac{1}{2}\sum_{i=1}^{n}(\mathbf{x}_i - \mu)^T C(\mathbf{x}_i - \mu) + const, \qquad (8.14)$$

where *const* is a constant that does not depend on the parameters μ and C. Without loss of generality, we will further assume that the data are *centered*, i.e. that $\mu = 0$, and hence the purpose is to estimate Σ, or its inverse, C. Thus, the second term in eq. 8.14 can be rewritten as $\frac{1}{2}\sum_{i=1}^{n}\mathbf{x}_i^T C\mathbf{x}_i = \frac{n}{2}tr(AC)$, the likelihood is

$$L(D) = \frac{n}{2}[\det(C) - tr(AC)] + const, \qquad (8.15)$$

thus the *log-likelihood maximization problem* can be now written as

$$(P1): \quad \max_{C \succ 0}\det(C) - tr(AC), \qquad (8.16)$$

where $A = \frac{1}{n}\sum_{i=1}^{n}\mathbf{x}_i^T\mathbf{x}_i$ is the *empirical covariance matrix*, or the maximum-likelihood estimate of Σ. Note that the $C \succ 0$ constraint ensures that C is positive definite.

The solution to 8.14 is the maximum-likelihood estimator (MLE) $\hat{C} = A^{-1}$, which appears to be the simplest way to handle the covariance-selection problem. However, the maximum-likelihood approach has several drawbacks. First, the inverse of the empirical covariance matrix, A^{-1}, may not even exist when the number of variables exceeds the number of samples, i.e. $p > n$. Second, even if A^{-1} exists (e.g., for $n \geq p$), it does not typically contain any zero elements, even when the number of samples is quite large[2]. Therefore, in order to learn the structure of a

---

[2]This may appear surprising at first, since A^{-1} is a *consistent* maximum-likelihood estimate of the true inverse-covariance matrix C, i.e. it converges to C with $n \to \infty$. However, consistency means convergence *in probability*, which means that, for each c_{ij}, and for all $\epsilon > 0$, $\lim_{n\to\infty}\Pr\left(|a_{ij} - c_{ij}| \geq \varepsilon\right) = 0$, i.e., with high probability, the estimates will be close to the true c_{ij} when the number of samples increases. Still, none of them have to be exact zero if the true c_{ij}, even for large n; indeed, this behavior is observed in various simulated experiments where the number of samples can be increased.

Gaussian MRF, i.e. to recover zeros of the concentration matrix, one must include an explicit sparsity-enforcing constraint into the maximum-likelihood formulation.

The *sparse inverse covariance selection problem* is to find the maximum-likelihood model with a constraint on the number of parameters (i.e., small l_0-norm of C). In general, this is an intractable combinatorial problem. Early approaches used greedy forward or backward search that required $O(p^2)$ maximum-likelihood-estimation (MLE) fits for different models in order to add (delete) an edge (Lauritzen, 1996), where p is the number of variables. This approach does not scale well with the number of variables[3]; moreover, as it was mentioned above, the existence of MLE for C is not even guaranteed when the number of variables exceeds the number of observations (Buhl, 1993).

Recently, however, an alternative approximation approach to the above problem was suggested in (Yuan and Lin, 2007; Banerjee et al., 2008) that replaces the intractable l_0 constraint with its l_1-relaxation, known to enforce sparsity, and yields a convex optimization problem that can be solved efficiently. In the subsequent sections, we discuss the problem formulation and optimization methods for this problem.

8.4.1   Sparse Inverse Covariance Selection Problem

Similarly to sparse regression, a common approach to enforcing sparsity in C is to impose Laplace priors $p(C_{ij}) = \frac{\lambda_{ij}}{2} e^{-\lambda_{ij}|C_{ij}|}$, with a common parameter $\lambda > 0$, on the elements of C, which is equivalent to adding the following l_1-norm penalty on the log-likelihood function in eq. 8.15:

$$L_{l_1}(D) = \frac{n}{2}[\ln \det(C) - tr(AC)] - \lambda ||C||_1, \qquad (8.17)$$

where the l_1-norm of C is simply the vector-norm $||C||_1 = \sum_{i,j} |C_{ij}|$. Generally, we assume that $\lambda \geq 0$, where the case of $\lambda = 0$ corresponds to the standard maximum-likelihood formulation. Then the joint log-likelihood maximization problem, first suggested in (Banerjee et al., 2006), is given by

$$\max_{C \succ 0} \ln \det(C) - tr(AC) - \rho ||C||_1, \qquad (8.18)$$

where $\rho = \frac{2}{n} \lambda$. Also, a more general assumption can be made about $p(C)$, allowing different elements of the matrix C to have different parameters λ_{ij} in the corresponding Laplace priors. This leads to the following formulation (e.g., in (Duchi et al., 2008; Scheinberg and Rish, 2010)):

$$\max_{C \succ 0} \ln \det(C) - tr(AC) - \sum_{ij} \rho_{ij}|C_{ij}|, \qquad (8.19)$$

which obviously reduces to the problem in 8.18 when $\rho_{ij} = \rho \;\; \forall i, j \in \{1, ..., p\}$. Herein, we will focus on the simpler formulation in the eq. 8.18, although the

---

[3]E.g., (Meinshausen and Bühlmann, 2006) reported difficulties when running "the forward selection MLE for more than thirty nodes in the graph".

approaches discussed below can be easily extended to the more general case in eq. 8.19.

For every given $\rho > 0$, the problem in eq. 8.18 is convex and has a unique solution (Banerjee et al., 2006, 2008). It can be solved in polynomial time by standard interior point methods (IPMs). For example, (Yuan and Lin, 2007) used the interior point method for the maxdet problem proposed by (Vandenberghe et al., 1998). However, the computational complexity of finding a solution within ϵ from the optimal is $O(p^6 \log(1/\epsilon))$, which becomes infeasible even for medium-sized problems with the number of variables on the order of hundreds. Moreover, interior point methods do not typically produce solutions that contain exact zeros, and thus require thresholding the elements of the matrix, potentially introducing inaccuracies in the zero-pattern recovery process.

As an alternative to IPMs, various efficient approaches were developed recently for the problem in eq. 8.18; we will discuss some of those approaches in the next section. Many of those approaches (e.g., (Banerjee et al., 2008; Friedman et al., 2007b; Duchi et al., 2008; Lu, 2009)) are focusing on solving the corresponding *dual problem*:

$$\max_{W \succ 0} \{\ln \det(W) : \|W - A\|_\infty \leq \rho\}, \tag{8.20}$$

where $\|X\|_\infty = \max_{i,j} |X_{ij}|$. Note that while both primal and dual problems are convex, the dual is also smooth, unlike the primal. The optimality conditions for this pair of primal and dual problems imply that $W_{ij} - A_{ij} = \rho_{ij}$ if $C_{ij} > 0$ and $W_{ij} - A_{ij} = -\rho_{ij}$ if $C_{ij} < 0$, and that $W = C^{-1}$, i.e. the solution of the dual gives an estimate $\hat{\Sigma}$ of the *covariance* matrix, while the primal problem obtains the *inverse covariance* estimate $\hat{\Sigma}^{-1}$. Note that the constraint $W \succ 0$ is enforced implicitly here, since $\ln \det(W) = -\infty$ when W is not positive-definite.

8.4.2 Optimization Approaches

In the past several years, the problem of learning the sparse inverse covariance matrix became a really active area of research that produced a wide variety of efficient algorithmic approaches, such as methods proposed by (Meinshausen and Bühlmann, 2006; Banerjee et al., 2006, 2008; Yuan and Lin, 2007; Friedman et al., 2007b; Rothman et al., 2008; Duchi et al., 2008; Marlin and Murphy, 2009; Schmidt et al., 2009; Honorio et al., 2009; Lu, 2009; Scheinberg and Rish, 2010; Scheinberg et al., 2010a), just to name a few. In this section, we will briefly describe some of those techniques.

8.4.2.1 Neighborhood selection via LASSO

One of the first approaches that applied the l_1-relaxation idea to learning zero-structure of a sparse inverse-covariance matrix was proposed by Meinshausen and Bühlmann (Meinshausen and Bühlmann, 2006). The idea was simple and elegant: for each variable X_i, learn its neighborhood (i.e., the nonzeros in the i-th row/column of the precision matrix) by solving the l_1-regularized linear regression (LASSO) problem with X_i as the target and the remaining variables as the regressors. A link between the two variables X_i and X_j is included into the Markov network (i.e., C_{ij} is nonzero), if the regression coefficient of either X_i on X_j, or X_j on X_i, is nonzero. (Alternatively, an AND-rule can be used (Meinshausen and Bühlmann, 2006)). As shown by (Meinshausen and Bühlmann, 2006), this approach can consistently estimate the network structure, i.e., zero-pattern of Σ^{-1}, provided that the sparsity parameter in the LASSO problems is selected properly, e.g., grows at a particular rate as p and n grow. This approach is simple and scalable to thousands of variables. Note, however, that is does not necessarily provide a consistent estimate of the actual *parameters* of the precision matrix, and its solutions may violate the symmetry and positive-definite constraints that the precision matrix must satisfy. As we discuss below, the neighborhood selection approach of (Meinshausen and Bühlmann, 2006) can be viewed as an approximation of the "exact" l_1-regularized maximum-likelihood problem in eq. 8.18. We will now review several state-of-the-art approaches to the latter problem.

8.4.2.2 Block-coordinate descent (BCD)

Among the first methods for the l_1-regularized likelihood maximization problem were two *block-coordinate descent (BCD)* algorithms, COVSEL by (Banerjee et al., 2006) and *glasso* by (Friedman et al., 2007b), applied to the *dual problem* in eq. 8.20. The idea of the BCD approach is outlined in Figure 8.4. Both BCD methods iteratively update one column/row of the matrix W (estimate of the covariance matrix) at a time, iterating until convergence, where a small enough duality gap is used as a convergence criterion (Banerjee et al., 2006, 2008):

$$tr(W^{-1}A) - p + \rho||W^{-1}||_1 \le \epsilon. \tag{8.21}$$

In step 2 of the BCD approach, we use $W_{/i}$ and $A_{/i}$ to denote the matrices obtained by removing the i-th column and the i-th row from the matrices W and A,

---

Block-Coordinate Descent (BCD)

1. **Initialize:** $W \leftarrow A + \rho I$

2. **For** $i = 1, ..., p$

    (a) Solve a box-constrained quadratic program in eq. 8.23:
    $\hat{y} = \arg\min_y \{y^T W_{/i}^{-1} y : ||y - A_i||_\infty \le \rho\}$

    (b) Update W: replace i-th column by $\hat{y}$ and i-th row by $\hat{y}^T$

3. **end for**

4. **If** convergence achieved, **then** return W; **otherwise** go to step 2.

FIGURE 8.4: Block-coordinate descent approach to solving the dual sparse inverse covariance problem.

respectively, while W_i and A_i denote the i-th columns of the corresponding matrices, with the diagonal elements removed. Thus, at each iteration, the empirical covariance matrix A and the current estimate of W are decomposed as follows:

$$W = \begin{pmatrix} W_{/i} & W_i \\ W_i^T & w_{ii} \end{pmatrix} \qquad A = \begin{pmatrix} A_{/i} & A_i \\ A_i^T & a_{ii} \end{pmatrix}. \tag{8.22}$$

Assuming that $W_{/i}$ is fixed, step 2 is to solve the optimization problem in eq. 8.20 with respect to the elements of the i-th column/row W_i (without the diagonal element). It can be shown that this problem reduces to the following box-constrained quadratic program:

$$\hat{y} = \arg\min_y \{y^T W_{/i}^{-1} y : ||y - A_i||_\infty \le \rho\}. \tag{8.23}$$

This subproblem is solved differently by COVSEL (Banerjee et al., 2006) and *glasso* (Friedman et al., 2007b). The COVSEL method solves this quadratic program using an interior-point approach. The overall computational complexity of the resulting algorithm is $O(Kp^4)$, where K is the total number of iterations, each iteration consisting of the full sweep over all columns in step 2. The *glasso* (Tibshirani, 1996) approach uses an alternative, and more efficient, way of solving the above quadratic subproblem. It solves instead the *dual* of the problem in eq. 8.23:

$$\min_x x^T W_{/i} x - A_i^T x + \rho||x||_1. \tag{8.24}$$

The above dual can be rewritten as a LASSO problem, using the notation $Q = (W_{/i})^{1/2}$ and $b = \frac{1}{2}Q^{-1}A_i$:

$$\min_x ||Qx - b||_2^2 + \rho||x||_1. \tag{8.25}$$

To solve the above LASSO problem, (Tibshirani, 1996) uses an efficient coordinate descent method suggested in (Friedman et al., 2007a). As a result, *glasso* is much faster than COVSEL empirically, and yields the overall (empirical) computational complexity of approximately $O(Kp^3)$, as discussed in (Duchi et al., 2008), although no explicit analysis was provided in the original paper by (Tibshirani, 1996).

Note that *glasso* clarifies the relationship between the l_1-regularized maximum-likelihood approach and the LASSO-based neighborhood selection approach of (Meinshausen and Bühlmann, 2006). Indeed, if instead of updating W we always use the empirical covariance matrix A, then $W_{/i} = A_{/i}$ and problem in 8.25 becomes equivalent to the l_1-penalized regression of X_i on the remaining variables. In general, however, $W_{/i} \ne A_{/i}$, except for the very first iteration. In other words, iteratively updating W in *glasso* takes into account the dependencies among the regressors, i.e. solves a sequence of *coupled* LASSO problems, while the approach of (Meinshausen and Bühlmann, 2006) treats them as completely independent subproblems. Thus, the method of (Meinshausen and Bühlmann, 2006) can be viewed as an approximation to the "exact" l_1-regularized maximum-likelihood.

Projected Gradient

1. $x \leftarrow x + \alpha \nabla f(x)$ (step of size α in the direction of gradient)

2. $x \leftarrow \Pi_S(x) = \arg\min_z \{||x - z||_2 : z \in S\}$ (project onto S)

3. **If** a convergence criterion is satisfied, **then** exit, **otherwise** go to step 1.

FIGURE 8.5: Projected gradient approach.

8.4.2.3 Projected gradient approach

In parallel with *glasso*, an efficient *projected gradient (PG)* approach was successfully applied to the dual problem in 8.20 by several authors (Duchi et al., 2008; Schmidt et al., 2008). Its time complexity is $O(Kp^3)$, i.e. the same order of magnitude as for *glasso*, but empirically it was shown to outperform *glasso* by the factor of two (Duchi et al., 2008). The high-level idea of the projected gradient approach is outlined in Figure 8.5, for a general optimization problem with a convex constraint set S:

$$\min_x \{f(x) : x \in S\}. \tag{8.26}$$

As its name suggests, the projected gradient approach iteratively updates x until convergence, making a step in the gradient direction, and projecting the result onto the constraint set S at each iteration (Bertsekas, 1976). First-order projected gradient algorithms are typically used in higher-dimensional problems where the second-order methods become intractable. (Duchi et al., 2008) applies the projected gradient approach to the dual problem in 8.20, where $x = W$, $f(W) = -\ln\det(W)$, and the convex set S is defined by the corresponding box-constraint in 8.20; similarly to COVSEL and *glasso*, the duality gap in eq. 8.21 is used as a convergence criterion. In (Schmidt et al., 2008), a modified version, called spectral projected gradient (SPG), is applied to the same problem. More recently, an even more efficient version of the projected gradient approach, called the *projected quasi-Newton (PQN)* method, was developed in (Schmidt and Murphy, 2010), and shown to outperform earlier PG approaches.

In the past couple years, project-gradient methods were considered the fastest state-of-art techniques available for the sparse inverse-covariance selection problem. However, most recently, even more efficient techniques were proposed, such as, for example, the alternating-linearization method (ALM) (Scheinberg et al., 2010a), described later in this section.

8.4.2.4 Greedy coordinate ascent on the primal problem

As we discussed above, multiple approaches to the sparse inverse-covariance selection problem focused on the dual formulation. Alternatively, several methods have been proposed recently for solving the primal problem directly, such as greedy coordinate descent (Scheinberg and Rish, 2010), block coordinate descent approaches (Sun et al., 2009; Honorio et al., 2009), and alternating linearization method (Scheinberg et al., 2010a). Herein, we discuss in more detail the method proposed in (Scheinberg and Rish, 2010), and called *SINCO* for Sparse INverse COvariance (Scheinberg and Rish, 2010). (We follow a more concise exposition of this method presented later in (Scheinberg and Ma, 2011).)

SINCO is a greedy coordinate ascent algorithm that optimizes only one diagonal or two symmetric off-diagonal elements of C at each iteration, unlike the COVSEL or *glasso*, which optimize one row (column) of the dual matrix W. More formally, at each iteration, the update step for the inverse covariance matrix C can be written as $C + \theta(e_i e_j^T + e_j e_i^T)$, where i and j are the indices of elements being changed, and θ is the step size. Thus, the objective function of the primal problem in 8.18 can be written as a function of θ, for a given C and (i, j):

$$f(\theta; C, i, j) = \ln \det(C + \theta e_i e_j^T + \theta e_j e_i^T) - tr(A(C + \theta e_i e_j^T + \theta e_j e_i^T)) -$$
$$\rho \| C + \theta e_i e_j^T + \theta e_j e_i \|_1. \tag{8.27}$$

In order to compute this function, (Scheinberg and Rish, 2010) use the following two properties of the determinant and its inverse: given a $p \times p$ matrix X and vectors $u, v \in \mathcal{R}^p$, we have

$$\det(X + uv^T) = \det(X)(1 + v^T X^{-1} u), \quad \text{and} \tag{8.28}$$
$$(X + uv^T)^{-1} = X^{-1} - X^{-1} uv^T X^{-1} / (1 + v^T X^{-1} u), \tag{8.29}$$

where the second equation is known as the Sherman-Morrison-Woodbury formula. At each iteration, SINCO greedily selects a pair (i, j) that yields the best improvement in the objective function and the step size θ. It then replaces the current estimate of C by $C + \theta(e_i e_j^T + e_j e_i^T)$, and updates $W = C^{-1}$ using the Sherman-Morrison-Woodbury formula given above. Figure 8.6 presents a high-level overview of the SINCO method. The key observation is that, given the matrix $W = C^{-1}$, the exact line search that optimizes $f(\theta)$ along the direction $e_i e_j^T + e_j e_i^T$ reduces to a solution of a quadratic equation, and thus takes a constant number of operations. Moreover, given the starting objective value, the new function value on each step can be computed in a constant number of steps. Thus, the for-loop over all (i, j) pairs takes $O(p^2)$ time. Updating the dual matrix $W = C^{-1}$ also takes $O(p^2)$ time, as it was shown in (Scheinberg and Rish, 2010), and thus each iteration takes $O(p^2)$ time. The steps of the SINCO algorithm are well-defined, i.e., the quadratic equation always yields the maximum of the function f along the chosen direction. The algorithm converges to the unique optimal solution of 8.18 and, more generally, of the problem in 8.19, in case of varying sparsity parameters (Scheinberg and Rish, 2010).

One of the main advantages of the SINCO approach is that it naturally preserves the sparsity of the solution and tends to avoid introducing unnecessary (small) nonzero elements. In practice, this often leads to lower false-positive error rates when compared to other approaches, such as COVSEL or *glasso*, while preserving the same false-negative rates, especially on very sparse problems, as discussed

---

SINCO

1. **Initialize:** $C = I$, $W = I$, $k = 0$

2. Next iteration: $k = k + 1$, $\theta_k = 0$, $i^k = 1$, $j^k = 1$, $f = f(\theta_k; C, i, j)$

3. **For** $i = 1, ..., p$, $j = 1, ..., p$ (choose best (i, j))

    (a) $\theta_{ij} = \arg\max_\theta f(\theta)$, where $f(\theta)$ is given in eq. 8.27
    (b) **If** $f(\theta_{ij}) > f(\theta^k)$, **then** $\theta_k = \theta_{ij}$, $i^k = i$, $j^k = j$

    end for

4. Update $C = C + \theta e_i e_j^T + \theta e_j e_i^T$ and $W = C^{-1}$

5. **If** $f(\theta^k) - f > \epsilon$, **then** go to step 2, **otherwise** return C and W.

FIGURE 8.6: SINCO - a greedy coordinate ascent approach to the (primal) sparse inverse covariance problem.

in (Scheinberg and Rish, 2010). Note that, although the current state-of-the-art algorithms for the above problem are converging to the same optimal solution in the limit, the near-optimal solutions obtained after any fixed number of iterations can be different structure-wise, even though they reach similar accuracy in the objective function reconstruction. Indeed, it is well-known that similar likelihoods can be obtained by two distributions with quite different structures due to multiple weak links. As to the l_1-norm regularization, although it often tends to enforce solution sparsity, it is still only an approximation to l_0 (i.e. a sparse solution may have the same l_1-norm as a much denser one). Adding l_1-norm penalty is only guaranteed to recover the "ground-truth" model under certain conditions on the data (that are not always satisfied in practice) and for certain asymptotic growth regimes of the regularization parameter, with growing number of samples n and dimensions p (with unknown constant). So the optimal solution, as well as near-solutions at given precision, could possibly include false positives, and one optimization method can potentially choose sparser near-solutions (at same precision) than another method.

Another advantage of SINCO is that each iteration takes only $O(p^2)$ operations, as opposed to to higher iteration costs of the previous methods. Note, however, that the overall number of iterations can be potentially higher for SINCO than, say, for block-coordinate methods updating a column/row at a time. Empirical results presented in (Scheinberg and Rish, 2010), and reproduced in Figure 8.7, demonstrate that while *glasso* is comparable to, or faster than, SINCO for a relatively small number of variables p, SINCO appears to have much better scaling when p increases (e.g., gets closer to 1000 variables), and can significantly outperform *glasso* (which, in turn, can be orders of magnitude faster than COVSEL). See (Scheinberg and Rish, 2010) for the details of the experiments. Moreover, SINCO's greedy approach

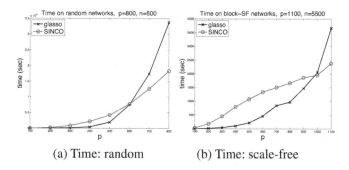

(a) Time: random   (b) Time: scale-free

FIGURE 8.7: CPU time comparison between SINCO and *glasso* on (a) random networks ($n = 500$, fixed range of ρ) and (b) scale-free networks that follow power-law distribution of node degrees (density 21%, n and ρ scaled by the same factor with p, $n = 500$ for $p = 100$).

introduces "important" nonzero elements into the inverse covariance matrix, in a manner similar to the path-construction process based on sequentially reducing the value of λ. SINCO can reproduce the regularization path behavior for a fixed (and sufficiently small) λ, without actually varying its value, but following instead the *greedy solution path*, i.e. sequentially introducing nonzero elements in a greedy way. Thus, SINCO can obtain any desired number of network links directly, without having to tune the λ parameter. This behavior is somewhat similar to LARS (Efron et al., 2004) for Lasso, however, unlike LARS, SINCO updates the coordinates which provide the best optimal function value improvement, rather than the largest gradient component.

Finally, while SINCO may not be the fastest method in a sequential setting (indeed, methods like the projected gradient of (Duchi et al., 2008) or the smooth optimization of (Lu, 2009) outperform *glasso*, which is comparable to SINCO), it is important to mention that SINCO lends itself to a *straightforward massive parallelization* at each iteration, due to the nature of its greedy steps, while none of its competitors seem to allow such an easy parallelization.

8.4.2.5 Alternating linearization method (ALM)

As we discussed above, coordinate-descent (CD) and block-coordinate descent (BCD) approaches are generally outperformed by gradient-based methods of (Duchi et al., 2008) and (Schmidt and Murphy, 2010). Another recently proposed gradient-based method for solving the primal problem in eq. 8.18, called *alternating linear minimization (ALM)* (Scheinberg et al., 2010a)[4], was shown to further outperform the projected-gradient methods of (Duchi et al., 2008) and (Schmidt and Murphy, 2010), and other state-of-the-art methods such as the smooth optimization approach of (Lu, 2009). Moreover, iteration complexity results are available for ALM, while no such results were provided for CD, BCDs and projected gradient methods; as shown in

---

[4]Not to be confused with ALM standing for Augmented Lagrange Multipliers method.

(Scheinberg et al., 2010a), ALM obtains an ϵ-optimal solution (i.e. a solution whose objective function is within ϵ from the optimum) in $O(1/\epsilon)$ iterations.

We now briefly discuss the main idea of the ALM method, which exploits the additive structure of the objective function in eq. 8.18. More specifically, given an optimization problem

$$\min_{\mathbf{x}} f(\mathbf{x}) + g(\mathbf{x}), \qquad (8.30)$$

where $f(\mathbf{x})$ and $g(\mathbf{x})$ are convex functions, one can separate these two function by introducing a new variable $\mathbf{y}$, which yields an equivalent problem:

$$\min_{\mathbf{x},\mathbf{y}} f(\mathbf{x}) + g(\mathbf{y}), \quad s.t. \ \mathbf{x} - \mathbf{y} = 0. \qquad (8.31)$$

An alternating direction augmented Lagrangian (ADAL) method can be now used for problem 8.31. The method iteratively updates each variable, $\mathbf{x}$ and $\mathbf{y}$, alternating between the two, using the following update rules:

$$\begin{cases} \mathbf{x}^{k+1} = \arg\min_{\mathbf{x}} L(\mathbf{x}, \mathbf{y}^k; \lambda^k) \\ \lambda_x^{k+1} = \lambda_y^k - (x^{k+1} - y^k)/\mu \\ \mathbf{y}^{k+1} = \arg\min_{\mathbf{y}} L(\mathbf{x}^{k+1}, \mathbf{y}; \lambda_x^{k+1}) \\ \lambda_y^{k+1} = \lambda_x^{k+1} - (x^{k+1} - y^{k+1})/\mu, \end{cases} \qquad (8.32)$$

where μ is a penalty parameter and $L(\mathbf{x}, \mathbf{y} : \lambda)$ is the augmented Lagrangian:

$$L(\mathbf{x}, \mathbf{y}; \lambda) = f(\mathbf{x}) + g(\mathbf{x}) - \lambda^T(\mathbf{x} - \mathbf{y}) + \frac{1}{2\mu}||\mathbf{x} - \mathbf{y}||^2. \qquad (8.33)$$

In case of smooth f and g, it can be shown that $\lambda_x^{k+1} = \nabla f(\mathbf{x}^{k+1})$ and $\lambda_y^{k+1} = -\nabla g(\mathbf{y}^{k+1})$, and thus at each iteration, updating $\mathbf{x}$ becomes equivalent to minimizing the sum of $f(\mathbf{x})$ and an approximation of $g(\mathbf{x})$ at the current $\mathbf{y}^k$, where the approximation is based on *linearization* of $g(x)$ and adding an "error" ("prox") term $\frac{1}{2\mu}||x - y^k||_2^2$. Similarly, updating $\mathbf{y}$ is equivalent to minimizing the sum of $g(y)$ and a linearlization of $f(\mathbf{y})$ at the current point $\mathbf{x}^{k+1}$ plus the corresponding prox term. Thus, the method is called *alternating linearization*.

Note that the primal problem in eq. 8.18 has the same decomposable form as discussed above, i.e. can be written as

$$\min_{C \succ 0} f(C) + g(C), \qquad (8.34)$$

where $f(C) = -\ln\det(C) + tr(AC)$ and $g(C) = \rho||C||_1$. Though these functions are not smooth, and the positive-definite constraint is imposed on C, a similar alternating linearization method can be derived for this problem, as shown in (Scheinberg et al., 2010a) (also, see (Scheinberg and Ma, 2011) for details).

8.4.3   Selecting Regularization Parameter

The previous section discussed several recently proposed methods for solving the sparse inverse-covariance problem. However, it left open an important question on how to choose the "right" level of sparsity, i.e. the regularization parameter λ. Clearly, the accuracy of the network structure reconstruction can be very sensitive to the choice of the regularization parameter, and it is a general consensus that optimal selection of λ in practical settings remains an open problem. Several approaches proposed in the literature include: (1) cross-validation, (2) theoretical derivations for asymptotic regime, (3) stability-selection (Meinshausen and Bühlmann, 2010), and (4) Bayesian treatment of the regularization parameter. We will briefly discuss these approaches, focusing in more detail on the Bayesian method suggested in (Asadi et al., 2009; Scheinberg et al., 2009).

At first, the standard cross-validation approach looks like the most natural approach. For each fixed λ, a model is learned on training data, and evaluated on a separate cross-validation dataset. The value of λ that yields the best likelihood on this dataset, i.e. best predicts the unseen data, is selected. Note, however, that optimizing the λ parameter for prediction may not necessarily lead to the best structure-reconstruction accuracy; moreover, it is well-known that probabilistic models having quite different graph structures may correspond to very similar distributions, especially in the presence of multiple "weak", or "noisy", edges (see, for example, (Beygelzimer and Rish, 2002)). Indeed, as it was observed empirically by several authors, λ selected by cross-validation with respect to likelihood is typically too small to provide an accurate structure recovery, i.e. it produces unnecessarily dense networks and thus yields high false-positive error rates. In fact, theoretical analysis provided by (Meinshausen and Bühlmann, 2006) for the neighborhood selection approach proves that the cross-validated λ *does not lead to consistent model selection*, since it tends to include too many noisy connections between the variables.

An alternative approach is analyze theoretically conditions on the regularization parameter that yield accurate structure recovery. However, most of the existing approaches (see, e.g., (Meinshausen and Bühlmann, 2006; Banerjee et al., 2008; Ravikumar et al., 2009)) focus on asymptotic regime, suggesting sufficient conditions for the growth rate of λ in order to guarantee a consistent estimate of the network structure as the number of samples n and the number of dimensions p grow. As noted in (Meinshausen and Bühlmann, 2006), "such asymptotic considerations give little advice on how to choose a specific penalty parameter for a given problem". As a proxy for "best" λ, Meinshausen and Bühlmann (2006) in their neighborhood-selection approach provide λ that allows a consistent recovery of sparse structure of the *covariance* rather than the *inverse covariance* matrix, i.e. the recovery of marginal independencies between i-th and j-th variables, rather than conditional independencies given the rest of the variables. A similar approach is used in (Banerjee et al., 2008) to derive λ for the optimization problem in eq. 8.17. In practice, however, this may give too high values of λ, i.e. sparsify the structure too much, missing connections among variables and thus leading to high false-negative rates. We will present empirical results that confirm this tendency empirically.

A recently proposed *stability selection* approach of (Meinshausen and Bühlmann, 2010) proposes a method for improving stability of sparse solutions to the choice of λ parameter in several sparse optimization problems, including both LASSO and sparse inverse-covariance selection. This approach, similar in spirit to BOLASSO of

(Bach, 2008a), randomly samples subsets of a given datasets, and solves the sparse recovery problem on each of them. A variable, such as an element of the inverse covariance matrix (i.e., a link in the graph), is included in the final model only if it appears in a sufficiently high fraction of models learned on those data subsets. As demonstrated by both (Meinshausen and Bühlmann, 2010) and (Bach, 2008a), this approach eliminates high sensitivity of sparse solutions to the choice of the regularization parameter. However, stability selection is computationally expensive as it requires to solve the same optimization problem on multiple subsets of the data.

Yet another alternative is to apply a Bayesian approach, treating λ as a random variable with some prior probability density $p(\lambda)$. Herein, we focus on a simple alternating-maximization method of (Asadi et al., 2009) for finding the maximum a posteriory probability (MAP) estimate of λ and C. As demonstrated empirically in (Asadi et al., 2009) and subsequently in (Scheinberg et al., 2009, 2010b), this method tends to produce a more balanced trade-off between false-positive and false-negative errors, the "overly inclusive" cross-validation parameter and "overly exclusive" theoretically derived parameter of (Banerjee et al., 2008). The joint distribution over C, λ, and $\mathbf{X}$ factorizes as follows: $p(\mathbf{X}, C, \lambda) = p(X|C)p(C|\lambda)p(\lambda)$, where $p(X|C)$ is a multivariate Gaussian distribution with zero mean and covariance C^{-1}, and $p(C|\lambda)$ is Laplace prior $p(C_{ij}) = \frac{\lambda_{ij}}{2}e^{-\lambda_{ij}|C_{ij}|}$, as discussed above. (For simplicity, we will again assume a common parameter $\lambda > 0$ for all elements C_{ij} of C, although the approach (Asadi et al., 2009) is easily extendable to the general case of varying sparsity parameters, as shown in (Scheinberg et al., 2009, 2010b)). Then the MAP estimate of λ and C is given by

$$\max_{C \succ 0, \lambda} \ln p(C, \lambda | \mathbf{X}) = \max_{C \succ 0, \lambda} \ln p(\mathbf{X}, C, \lambda) = \max_{C \succ 0, \lambda} \ln[p(X|C)p(C|\lambda)p(\lambda)].$$
(8.35)

This results in the following MAP problem, which essentially adds two extra terms to the earlier formulation given by eq. 8.17:

$$\max_{\lambda, C \succ 0} \frac{n}{2}[\ln \det(C) - tr(AC)] - \lambda ||C||_1 + p^2 \ln \frac{\lambda}{2} + \ln p(\lambda).$$

In (Asadi et al., 2009; Scheinberg et al., 2009, 2010b), several types of priors on λ are considered, including exponential, uniform (flat), and truncated Gaussian priors. The *uniform (flat) prior* puts equal weight on all values of $\lambda \in [0, \Lambda]$ (assuming sufficiently high Λ), thus the last term $\ln p(\lambda)$ in eq. 8.36 is effectively ignored. The *exponential prior* assumes that $p(\lambda) = be^{-b\lambda}$, yielding

$$\max_{\lambda, C \succ 0} \frac{n}{2}[\ln \det(C) - tr(AC)] + p^2 \ln \frac{\lambda}{2} - \lambda ||C||_1 - b\lambda.$$
(8.36)

Rather than taking a more expensive, fully Bayesian approach here and integrating out C in order to obtain the estimate of b, (Asadi et al., 2009) use a somewhat ad-hoc approximate estimate $b = ||A_r^{-1}||_1/(p^2 - 1)$, where $A_r = A + \epsilon I$ is the empirical covariance matrix (slightly regularized with small $\epsilon = 10^{-3}$ on the diagonal to obtain an invertible matrix when A is not invertible). The intuition behind such estimate

---

Alternating Maximization Algorithm for Regularization Parameter Selection

1. **Initialize** λ^1

2. Find $C(\lambda^k)$, $\phi(\lambda^k)$ and $\psi(\lambda^k)$

3. **If** $|p^2/\lambda - \|C(\lambda^k)\|_1 - b| < \epsilon$, **go to** step 6

4. $\lambda^{k+1} = p^2/(\|C(\lambda^k)\|_1 + b)$

5. Find $C(\lambda^{k+1})$ and $\psi(\lambda^{k+1})$
   if $\psi(\lambda^{k+1}) > \psi(\lambda^k)$ **go to** step 4
   else $\lambda^{k+1} = (\lambda^k + \lambda^{k+1})/2$, **go to** step 5

6. **Return** λ^k and $C(\lambda^k)$

FIGURE 8.8: Alternating maximization scheme for simultaneous selection of the regularization parameter λ and the inverse-covariance matrix C.

is that $b = 1/E(\lambda)$, and we set $E(\lambda)$ to be the solution of the above optimization problem with C fixed to its empirical estimate A_r^{-1}. Also, (Scheinberg et al., 2009, 2010b) consider the unit-variance Gaussian prior, truncated to exclude negative values of λ.

We will consider here the exponential prior and the corresponding optimization problem in eq. 8.36. The objective function is concave in C for any fixed λ but is not concave in C and λ jointly. Thus, we are looking for a local optimum. An alternating maximization approach proposed in (Asadi et al., 2009) solves the following problem, for each given fixed value of λ:

$$\phi(\lambda) = \max_C \frac{n}{2} \ln \det(C) - \frac{n}{2} tr(AC) - \lambda \|C\|_1 \qquad (8.37)$$

Given λ, this problem reduces to the standard sparse inverse-covariance problem, and thus has a unique maximizer $C(\lambda)$ (Banerjee et al., 2008). We now consider the following optimization problem:

$$\max_\lambda \psi(\lambda) = \max_\lambda \phi(\lambda) + p^2 \ln \lambda - b\lambda. \qquad (8.38)$$

Clearly, the optimal solution to this problem is also optimal for problem 8.36. Figure 8.8 shows a simple optimization scheme for problem 8.38 suggested in (Asadi et al., 2009). This scheme uses line search along the direction of the derivatives and will converge to the local maximum (if one exists) as long as some sufficient increase condition (such as Armijo rule Nocedal and Wright (2006)) is applied in Step 4. Step 1 can be performed by any convex optimization method designed to solve problem 8.37.

Empirical evaluation presented in (Asadi et al., 2009; Scheinberg et al., 2009, 2010b) demonstrated advantages of the above MAP approach for λ selection over

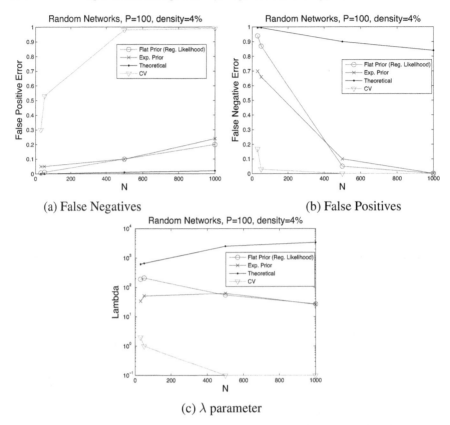

(a) False Negatives   (b) False Positives

(c) λ parameter

FIGURE 8.9: MAP approach versus cross-validated and theoretical methods for regularization parameter selection, on sparse random networks (4% density).

both cross-validation-based and theoretically derived λ. (In all experiments, *glasso* (Friedman et al., 2007b) method was used to solve the sparse inverse-covariance selection subproblem in the alternating minimization scheme). Figure 8.9 reproduces results on randomly generated synthetic problems. The "ground-truth" random inverse-covariance matrices of two levels of sparsity were used: a very sparse one, with only 4% (off-diagonal) nonzero elements, and a relatively dense one, with 52% (off-diagonal) nonzero elements. $n = 30, 50, 500, and 1000$ instances were sampled from the corresponding multivariate Gaussian distribution over $p = 100$ variables. The structure-learning performance of the Bayesian λ was compared against the two other alternatives: (1) λ selected by cross-validation and (2) theoretically derived λ from (Banerjee et al., 2008). Figure 8.9 shows the results on a sparse random network (4% link density); very similar results were also obtained for dense (52% link density) random matrices. We can clearly see that: (1) cross-validated λ (green) over-fits dramatically, producing an almost complete matrix (almost 100% false-positive rate); (2) theoretically derived λ Banerjee et al. (2008) (shown in black) is too conservative: it misses almost all edges (has a very high false-negative rate); (3) prior-based approaches - flat prior (red) and exponential prior (blue) yield much more balanced trade-off between the two typos of errors.

8.5 Summary and Bibliographical Notes

Probabilistic graphical models are a popular research topic in statistics and machine learning, with a long history and extensive literature. There are many books providing comprehensive coverage of different aspects of graphical models, such as (Pearl, 1988; Whittaker, 1990; Lauritzen, 1996; Cox and Wermuth, 1996; Cowell et al., 1999; Pearl, 2000; Edwards, 2000; Jordan, 2000; Koller and Friedman, 2009). This chapter only covers some aspects of undirected graphical models, focusing on recent advances in learning Gaussian MRFs with sparse structure, while several other importnant recent developments in sparse graphical model learning remain out of the scope of this book. For example, sparse learning with discrete-valued MRFs was addressed in (Wainwright et al., 2007) by extending the LASSO-based neighborhood-selection approach of (Meinshausen and Bühlmann, 2006) to binary variables via sparse logistic regression. Asymptotic consistency analysis of sparse Gaussian MRFs was given in (Ravikumar et al., 2009). (Lee et al., 2006b) learn MRFs using clique selection heuristic and approximate inference. Learning sparse *directed* networks, such as Bayesian networks, for both discrete and continuous variables, is addressed, for example, in (Schmidt et al., 2007; Huang et al., 2013; Xiang and Kim, 2013). (Lin et al., 2009) propose an alternative approach based on ensemble-of-trees that is shown to sometimes outperform l_1-regularization approaches of (Banerjee et al., 2008) and (Wainwright et al., 2007), (Schmidt and Murphy, 2010) propose a method for learning log-linear models with higher-order (beyond pairwise) potentials; group-l_1 regularization with overlapping groups is used to enforce hierarchical structure over potentials. A recent doctoral thesis by (Schmidt, 2010) discusses several state-of-art optimization approaches to learning both directed and undirected sparse graphical models. Finally, various algorithms for the sparse inverse-covariance estimation problem were proposed recently, including some of the methods discussed above, such as SINCO, a greedy coordinate-descent approach of (Scheinberg and Rish, 2010), alternating linearization method (ALM) (Scheinberg et al., 2010a), projected gradient (Duchi et al., 2008), block coordinate descent approaches (Sun et al., 2009; Honorio et al., 2009), variable-selection (group-sparsity) GMRF learning method (Honorio et al., 2012), as well as multiple other techniques (Marlin and Murphy, 2009; Schmidt et al., 2009; Lu, 2009; Yuan, 2010; Cai et al., 2011; Olsen et al., 2012; Kambadur and Lozano, 2013; Honorio and Jaakkola, 2013; Hsieh et al., 2013).

Chapter 9

Sparse Matrix Factorization: Dictionary Learning and Beyond

In this chapter, we will focus on sparsity in the context of matrix factorization, and consider an approximation $Y \approx AX$ of an observed matrix Y by a product of two unobserved matrices, A and X. Common data-analysis methods such as Principal Component Analysis (PCA) and similar techniques can be formulated as matrix factorization problems. This formulation is also central to a highly popular and promising research direction in signal processing and statistics known as *dictionary learning*, or *sparse coding* (Olshausen and Field, 1996), discussed in more detail below.

Note that the standard sparse signal recovery setting considered so far assumes that the design matrix A is known in advance. For example, Fourier transform matrices and random matrices were discussed earlier in this book as two "classical" examples of dictionaries that generated a wealth of theoretical results on sparse recovery. Various other dictionaries, such as wavelets, curvelets, contourlets, and other transforms, were also proposed recently, especially in the context of image processing (Elad, 2010).

However, a fixed dictionary may not necessarily be the best match for a particular type of signals, since a given basis (columns of A, or dictionary elements) may not yield a sufficiently sparse representation of such signals. Thus, a promising alternative approach that became popular in past years is to *learn* a dictionary that allows for a sparse representation, given a training set of observed signal samples. Figure 9.1 illustrates this approach: given a data matrix Y, where each column represents an observed signal (sample), we want to find the design matrix, or dictionary, A, as well as a sparse representation of each observed signal in that dictionary, corresponding to the sparse columns of the matrix X.

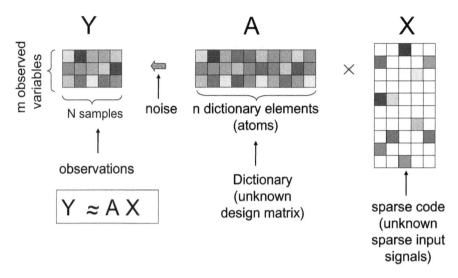

FIGURE 9.1: Dictionary learning, or sparse coding, as matrix factorization; note that the "code" matrix X is assumed to be sparse.

An alternative approach in sparse matrix factorization is to search for sparse dictionary elements (columns of A) rather than for sparse representations (columns of X). The motivation for such formulation, used in sparse PCA methods reviewed later in this chapter, is to improve the interpretability of dictionary elements, or components, since sparse components will help to identify small subsets of the "most important" input variables.

In the next section, we discuss dictionary learning and present several commonly used algorithms for this problem, using both l_0- and l_1-norm constraints to enforce sparsity. Next, we review the sparse PCA problem and methods. We also discuss other examples of sparse matrix factorizations, such as the sparse Nonnegative Matrix Factorization (NMF) approach of (Hoyer, 2004), in the context of the blind-source separation problem and its applications to the computer network diagnosis (Chandalia and Rish, 2007).

9.1   Dictionary Learning

The problem of *dictionary learning*, also known as *sparse coding*, was initially studied by (Olshausen and Field, 1996, 1997) in the context of neuroscience. Dictionary learning was proposed as a model for the evolutionary process that gave rise to the existing population of simple cells in the mammalian visual cortex. The approach of (Olshausen and Field, 1996) was further developed and extended by multiple researchers, including (Lewicki and Olshausen, 1999; Lewicki and Sejnowski, 2000; Engan et al., 1999; Kreutz-Delgado et al., 2003; Lesage et al., 2005; Elad and Aharon, 2006; Aharon et al., 2006a,b; Yaghoobi et al., 2009; Skretting and Engan, 2010; Mairal et al., 2009, 2010; Tosic and Frossard, 2011), and many others. Various algorithms for dictionary learning have been proposed, including K-SVD of (Aharon et al., 2006a), the Method of Optimal Directions (MOD) of (Engan et al., 1999), the online dictionary learning approach of (Mairal et al., 2009, 2010), to name a few. In the past several years, dictionary learning and sparse coding became popular research topics in the machine-learning community, introducing novel methods and applications of this framework (Lee et al., 2006a; Gregor and LeCun, 2010).

9.1.1 Problem Formulation

We now formally state the dictionary learning, or sparse coding, problem. Let $\mathbf{Y}$ be an $m \times N$ matrix, where $m < N$, and the i-th column, or sample, $\mathbf{y}_i$, is a vector of observations obtained using linear projections specified by some unknown $m \times n$ matrix $\mathbf{A}$, of the corresponding sparse column-vector $\mathbf{x}_i$ of the (unobserved) $n \times N$ matrix $\mathbf{X}$. For example, if the columns of $\mathbf{Y}$ are (vectorized) images, such as fMRI scans of a brain, then the columns of $\mathbf{A}$ are dictionary elements, or atoms (i.e., some "elementary" images, for example, corresponding to particular brain areas known to be activated by specific tasks and/or stimuli), and the columns of $\mathbf{X}$ correspond to sparse codes needed to represent each image using the dictionary (i.e., one can hypothesize that the brain activation observed in a given fMRI scan can be represented as a weighted linear superposition of a relatively small number of active brain areas, out of a potentially large number of such areas).

The ultimate sparse-coding objective is to find both $\mathbf{A}$ and $\mathbf{X}$ that yield the sparsest representation of the data $\mathbf{Y}$, subject to some acceptable approximation error ϵ:

$$(D_0^\epsilon): \quad \min_{\mathbf{A},\mathbf{X}} \sum_{i=1}^{N} ||\mathbf{x}_i||_0 \text{ subject to } ||\mathbf{Y} - \mathbf{A}\mathbf{X}||_2 \leq \epsilon. \qquad (9.1)$$

Note that this problem formulation looks very similar to the classical sparse signal recovery problem P_0^ϵ, only with two modifications: (1) dictionary $\mathbf{A}$ is now included as an unknown variable that we must optimize over, and (2) there are M, rather than just one, observed samples and the corresponding sparse signals, or sparse codes. Similarly to the sparse signal recovery problem, the question is whether the above problem can have a unique solution, assuming that obvious non-uniqueness issues due to scaling and permutation of the dictionary elements are taken care of, e.g., by normalizing the columns of $\mathbf{A}$ and fixing their ordering. In (Aharon et al., 2006b), it was shown that, in the case of $\epsilon = 0$, the answer to the above question is positive, provided that the matrices $\mathbf{Y}$, $\mathbf{A}$, and $\mathbf{X}$ satisfy certain conditions; namely, the set of samples $\mathbf{Y}$ must be "sufficiently diverse" and these samples should allow for a "sufficiently sparse" representation (e.g., using less than $spark(\mathbf{A})/2$ elements), in some dictionary $\mathbf{A}$. Then such dictionary $\mathbf{A}$ is unique, up to re-scaling and permutation of the columns.

As usual, there are also two alternative ways of formulating the above constrained optimization problem, i.e., by reversing the roles of the objective and the constraint:

$$(D_0^t): \quad \min_{\mathbf{A},\mathbf{X}} ||\mathbf{Y} - \mathbf{A}\mathbf{X}||_2^2 \text{ subject to } ||\mathbf{X}(i,:)||_0 \leq k, \ 1 \leq i \leq N,$$

for some k that corresponds to the above ϵ, or by using the Lagrangian relaxation:

$$(D_0^\lambda): \quad \min_{\mathbf{A},\mathbf{X}} ||\mathbf{Y} - \mathbf{A}\mathbf{X}||_2^2 + \lambda \sum_{i=1}^{N} ||\mathbf{X}_{i,:}||_0.$$

Clearly, the computational complexity of dictionary learning is at least as high as the complexity of the original (NP-hard) l_0-norm minimization problem. Thus, the l_1-norm relaxation can be applied, as before, to at least convexify the subproblem concerned with optimizing over the $\mathbf{X}$ matrix. Also, it is common to constrain the norm of the dictionary elements (e.g., by unit norm), in order to avoid arbitrarily large values of $\mathbf{A}$ elements (and, correspondingly, infinitesimal values of $\mathbf{X}$ entries) during the optimization process, leading to the following formulation (Mairal et al., 2010):

$$(D_1^\lambda): \quad \min_{\mathbf{A},\mathbf{X}} ||\mathbf{Y} - \mathbf{A}\mathbf{X}||_2^2 + \lambda \sum_{i=1}^{N} ||\mathbf{X}_{i,:}||_1$$
$$\text{subject to } ||\mathbf{A}_{:,j}||_2 \leq 1, \ \forall \, j = 1, ..., n. \qquad (9.2)$$

Given a fixed $\mathbf{A}$, the optimization over $\mathbf{X}$ is now convex; note, however, that the joint optimization over both $\mathbf{A}$ and $\mathbf{X}$ still remains non-convex.

9.1.2 Algorithms for Dictionary Learning

A common approach to non-convex matrix-factorization problems is to use the *alternating-minimization*, or the *block-coordinate descent (BCD)* approach, which iterates until convergence between the two optimization steps: (1) optimizing with respect to $\mathbf{X}$, given a fixed $\mathbf{A}$, and (2) optimizing with respect to $\mathbf{A}$, given a fixed $\mathbf{X}$. In the next section, we consider examples of commonly used algorithms for dictionary learning.

Figure 9.2 presents a simple alternating-maximization approach known as Method of Optimal Directions (MOD), introduced by (Engan et al., 1999); also, see

---

Method of Optimal Directions (MOD)

Input: $m \times N$ matrix of samples $\mathbf{Y}$, sparsity level k, precision ϵ.
Initialize: generate a random $m \times n$ dictionary matrix $\mathbf{A}$, or construct $\mathbf{A}$ using n randomly selected samples (columns) from $\mathbf{Y}$. Normalize $\mathbf{A}$.
Alternating minimization loop:

1. **Sparse coding**: for each $1 \leq i \leq N$, solve (using, e.g., MP or OMP)

$$\mathbf{x}_i = \arg\min_{\mathbf{x}} ||\mathbf{y}_i - \mathbf{A}\mathbf{x}||_2^2 \text{ subject to } ||\mathbf{x}||_0 \leq k$$

   to obtain the i-th sparse column of $\mathbf{X}$.

2. **Dictionary Update**:

$$\mathbf{A} = \arg\min_{\hat{\mathbf{A}}} ||\mathbf{Y} - \hat{\mathbf{A}}\mathbf{X}||_2^2 = \mathbf{Y}\mathbf{X}^T(\mathbf{X}\mathbf{X}^T)^{-1}.$$

3. **Stopping criterion**: If the change in the approximation error $||\mathbf{Y} - \mathbf{A}\mathbf{X}||_2^2$ since the last iteration is less than ϵ, then exit and return the current $\mathbf{A}$ and $\mathbf{X}$, otherwise go to step 1.

FIGURE 9.2: Method of Optimal Directions (MOD) for dictionary learning.

(Elad, 2010). Note that the dictionary can be initialized in various ways, e.g., generated randomly or constructed as a random subset of observed samples. It is also normalized to avoid scaling issues. Then, given the current dictionary $\mathbf{A}$, in step 1, we solve the standard l_0-norm sparse recovery problem for each sample (column i) in $\mathbf{Y}$, using, for example, greedy matching pursuit methods discussed above, such as MP or OMP. Once the collection of sparse codes for all samples, i.e., the matrix $\mathbf{X}$, is computed, we perform the dictionary update step (step 2 in the algorithm) using the least-squares minimization, and also obtain the current approximation error $||\mathbf{Y} - \mathbf{A}\mathbf{X}||_2^2$. If the decrease in the error is sufficiently small, we declare that the algorithm has converged, and return the computed dictionary $\mathbf{A}$ and the sparse code $\mathbf{X}$; otherwise, we continue alternating iterations.

Another well-known dictionary-learning approach, K-SVD, was introduced by (Aharon et al., 2006a) several years after MOD. K-SVD employs a different dictionary-update step, where each column of the dictionary is updated separately; empirical results demonstrate some improvements in performance of K-SVD over MOD (Elad, 2010). Moreover, in the past several years, various novel techniques for dictionary learning were developed.

One of the notable advances was the *online* dictionary learning introduction by (Mairal et al., 2009, 2010). Unlike the *batch* approaches discussed so far, which attempt to learn a dictionary directly from the full data set, online approaches process training samples incrementally, one training sample (or a small batch of training

---

Online dictionary learning

Input: a sequence of input samples $\mathbf{y} \in \mathbb{R}^m$, regularization parameter λ, initial dictionary $\mathbf{A}_0 \in \mathbb{R}^{m \times n}$, number of iterations T, threshold ϵ.
Initialize: $\mathbf{U}_0 \in \mathbb{R}^{n \times n} \leftarrow 0$, $\mathbf{V}_0 \in \mathbb{R}^{m \times n} \leftarrow 0$.

For $i = 1$ to T

    1. obtain next input sample $\mathbf{y}_i$

    2. **Sparse coding**: compute sparse code $\mathbf{x}_i$:

$$\mathbf{x}_i = \arg \min_{\mathbf{x} \in \mathbb{R}} \frac{1}{2} ||\mathbf{y}_i - \mathbf{A}_{i-1}\mathbf{x}||_2^2 + \lambda ||\mathbf{x}||_1$$

    3. $\mathbf{U}_i \leftarrow \mathbf{U}_{i-1} + \mathbf{x}_i \mathbf{x}_i^T, \quad \mathbf{V}_i \leftarrow \mathbf{V}_{i-1} + \mathbf{y}_i \mathbf{x}_i^T$.

    4. **Dictionary update**: use the BCD algorithm (Figure 9.4), with the input parameters $\mathbf{A}_{i-1}$, $\mathbf{U}_i$ and $\mathbf{V}_i$, to update the current dictionary by solving

$$\mathbf{A} = \arg \min_{\mathbf{A} \in S} \frac{1}{i} \sum_{j=1}^{i} (\frac{1}{2} ||\mathbf{y}_i - \mathbf{A}\mathbf{x}_i||_2^2 + \lambda ||\mathbf{x}_i||_1) =$$

$$\arg \min_{\mathbf{A} \in S} \frac{1}{i} (\frac{1}{2} Tr(\mathbf{A}^T \mathbf{A} \mathbf{U}_i) - Tr(\mathbf{A}^T \mathbf{V}_i),$$

where $S = \{\mathbf{A} = [\mathbf{a}_1, ..., \mathbf{a}_n] \in \mathbb{R}^{m \times n} \text{ s.t. } ||\mathbf{a}_j||_2 \leq 1, \forall j = 1, ..., n \}$.

Return $\mathbf{A}_i$.

FIGURE 9.3: Online dictionary-learning algorithm of (Mairal et al., 2009).

samples) at a time, similarly to stochastic gradient descent. At each iteration, a sparse code is computed for the new sample, and the dictionary is updated accordingly. One of the key motivations behind an online approach is that it scales much better than the batch techniques in applications with a very large number of training samples, for example, in image and video processing, where dictionaries are often learned on small image patches, and the number of such patches can be on the order of several millions. Moreover, as shown in (Mairal et al., 2009, 2010), the proposed algorithm provably converges, and tends to outperform batch approaches both in terms of the speed and the quality of the dictionary learned, on large as well as on small datasets.

    Figure 9.3 shows the details of the online dictionary-learning method proposed by (Mairal et al., 2010). Note that unlike the MOD algorithm presented above, this method considers the l_1-regularized problem formulation stated in eq. 9.2. At each iteration i of the algorithm, the next sample, $\mathbf{y}_i$, is drawn, then the sparse code for

---

Block Coordinate Descent (BCD) for Dictionary Update

Input: initial dictionary $\mathbf{A} \in \mathbb{R}^{m \times n}$ (for warm restart); auxiliary matrices $\mathbf{U} = [\mathbf{u}_1, ..., \mathbf{u}_n] \in \mathbb{R}^{n \times n}$, $\mathbf{V} = [\mathbf{v}_1, ..., \mathbf{v}_n] \in \mathbb{R}^{m \times n}$, threshold ϵ.
Repeat until convergence of $\mathbf{A}$:

   1. for each $j = 1$ to n, update j-th dictionary element:

$$\mathbf{a}_j \leftarrow \frac{1}{\max(||\mathbf{z}_j||_2, 1)}\mathbf{z}_j, \ where \ \mathbf{z}_j \leftarrow \frac{1}{\mathbf{u}_{jj}}(\mathbf{v}_j - \mathbf{A}\mathbf{u}_j) + \mathbf{a}_j,$$

   2. If the change in $||\mathbf{A}||_2$ on the last two iterations is above ϵ, go to step 3.

Return learned dictionary $\mathbf{A}_i$.

FIGURE 9.4: Block coordinate descent (BCD) for dictionary update in the online dictionary-learning method of (Mairal et al., 2009).

this sample is computed, by solving the standard Lasso problem

$$\min_{\mathbf{x} \in \mathbb{R}} \frac{1}{2}||\mathbf{y}_i - \mathbf{A}_{i-1}\mathbf{x}||_2^2 + \lambda||\mathbf{x}||_1,$$

where $\mathbf{A}$ is the current dictionary. (Mairal et al., 2009, 2010) use the LARS algorithm for solving this problem, though, of course, any Lasso solver can be applied here. After the sparse code for the current sample is obtained, the algorithm updates the two auxiliary matrices, $\mathbf{U}$ and $\mathbf{V}$; these matrices are used later in the dictionary update step. Finally, the dictionary-update step computes the new dictionary by minimizing the following objective function $\hat{f}_i(\mathbf{A})$ over the subset S of dictionaries with norm-bounded columns/elements, using the block-coordinate descent algorithm with warm restarts, presented in Figure 9.4:

$$\hat{f}_i(\mathbf{A}) = \frac{1}{i}\sum_{j=1}^{i}(\frac{1}{2}||\mathbf{y}_i - \mathbf{A}\mathbf{x}_i||_2^2 + \lambda||\mathbf{x}_i||_1),$$

where each $\mathbf{x}_j$, $1 \leq j < i$ was computed at the j-th previous iteration. This function serves as a surrogate for the corresponding (batch) empirical loss function

$$f_i(\mathbf{A}) = \frac{1}{i}\sum_{j=1}^{i} L(\mathbf{y}_j, \mathbf{A}),$$

where

$$L(\mathbf{y}_j, \mathbf{A}) = \min_{\mathbf{x}}(\frac{1}{2}||\mathbf{y}_i - \mathbf{A}\mathbf{x}||_2^2 + \lambda||\mathbf{x}||_1).$$

In other words, the (batch) empirical loss function $f_i(\mathbf{A})$ is obtained assuming that the optimal sparse codes for all samples are computed at once, with respect to a fixed dictionary $\mathbf{A}$, while its online surrogate loss function $\hat{f}_i(\mathbf{A})$ uses the sparse codes from earlier iterations. The key theoretical contribution of (Mairal et al., 2009, 2010) shows that both $f_i(\mathbf{A}_i)$ and $\hat{f}_i(\mathbf{A}_t)$ converge almost surely to the same limit. Finally, (Mairal et al., 2010) present several augmentations of the baseline approach presented above that improve its efficiency, including rescaling the "past" data so that the new coefficients $\mathbf{x}_i$ have higher weights, using mini-batches instead of single samples at each iteration, deleting dictionary atoms that are used very infrequently, and so on.

9.2   Sparse PCA

9.2.1   Background

Principal component analysis (PCA) is a popular data-analysis and dimensionality-reduction tool with a long history dating back to 1901 (Pearson, 1901) and a wide range of applications in statistics, science, and engineering. PCA assumes as an input a set of data points in a high-dimensional space defined by a set of potentially correlated input variables, and applies an orthogonal transformation that maps those points to another space, defined by a (smaller or equal) set of uncorrelated new variables, called *principal components*. The objective of PCA is to reduce the dimensionality while preserving as much variability in the data as possible. In order to achieve this, principal components are defined as orthogonal directions that account for the maximum variance in the data; namely, the first principal component is the direction of the largest possible variance, and each subsequent component maximizes the remaining variance subject to the constraint of being uncorrelated with (i.e., orthogonal to) the previous components[1].

PCA can be considered from two alternative perspectives, commonly referred to in the literature as the *analysis* view and the *synthesis* view (Jenatton et al., 2010). The traditional analysis view assumes the sequential approach outlined above, which finds the principal components one at a time, iteratively alternating between the variance-maximization to find the next component, and transformation (deflation) of the current covariance matrix to eliminate the influence of the previous components. Specifically, let $m \times N$ matrix $\mathbf{Y}$ represent a data matrix containing N data points, or samples, as its columns; the rows correspond to m input variables, or dimensions. The rows of $\mathbf{Y}$, corresponding to the input variables, are assumed to be centered to have zero empirical mean. PCA finds a (norm-bounded) vector of *loadings* $\mathbf{a} \in \mathbb{R}^m$ for the first principal component, so that projecting the data samples on $\mathbf{a}$ yields a new, highest-variance one-dimensional dataset; in other words, PCA finds the set of *scores* $\mathbf{x}_{1i}, 1 \leq 1 \leq N$ of the first principal component $\mathbf{x}_1 = \mathbf{Y}^T \mathbf{a}$, that maximizes

---

[1] Note that in the case of Gaussian data, the orthogonality, or lack of correlation, constraint is sufficient to guarantee that components are independent, which is not necessarily true in the case of more general data distributions. In such cases, Independent Component Analysis (ICA) can be used to find *independent* (rather than principal) components.

the variance of the scores, $\sum_i^N (\mathbf{y}_i^T \mathbf{a})^2 = ||\mathbf{Y}^T \mathbf{a}||_2^2 = \mathbf{a}^T \mathbf{Y} \mathbf{Y}^T \mathbf{a}$, i.e.,

$$\mathbf{a} = \arg \max_{||\mathbf{a}||_2 \leq 1} \mathbf{a}^T \mathbf{C} \mathbf{a},$$

where $\mathbf{C} = \mathbf{Y} \mathbf{Y}^T$ is (proportional to) the empirical covariance matrix. The above problem turns out to be equivalent to finding the largest eigenvalue and the corresponding eigenvector of $\mathbf{C}$, and finding subsequent principal components also turns out to be equivalent to finding the remaining eigenvectors of $\mathbf{C}$.

An alternative view at PCA, i.e., the synthesis view, also sometimes referred to as *probabilistic PCA* (Tipping and Bishop, 1999), consists in finding an orthogonal set of new basis vectors, or dictionary elements (loadings), as discussed before, as columns of an $m \times k$ matrix $\mathbf{A}$, where k is the desired number of components, and the new representation of the data samples in this basis (i.e., the corresponding new coordinates as the projections on the new basis vectors), given by the columns of an $k \times N$ matrix $\mathbf{X}$, where the columns correspond to data samples represented in the new basis. In a high-dimensional setting where $m \geq N$, i.e., the number of variables is greater than the number of samples, it is common to search for only a few first components, i.e., to assume $k \ll m$. The matrices $\mathbf{A}$ and $\mathbf{X}$ are found by solving the matrix-factorization problem, as discussed above, that minimizes the data reconstruction error:

$$\min_{\mathbf{A},\mathbf{X}} ||\mathbf{Y} - \mathbf{A}\mathbf{X}||_2^2, \qquad (9.3)$$

where the columns of $\mathbf{A}$ are often assumed to have unit-bounded norm, $||\mathbf{a}_i||_2 \leq 1$, as discussed earlier in the dictionary-learning context. Note, however, that in the literature on the matrix-factorization (probabilistic) approach to PCA, the orthogonality constraint on the dictionary elements is often omitted; as a result, the solution vectors do not always coincide with the principal components, but rather span the same space as the principal components (Tipping and Bishop, 1999). Orthogonalization of such matrix-factorization solution will recover the principal components.

Note that the above matrix-factorization approach to finding the first k principal components is closely related to the singular value decomposition (SVD) of the data matrix. Namely, as mentioned above, let the rows of $\mathbf{Y}$ (i.e., the input variables) be centered to have zero means, and let the rank of $\mathbf{Y}$ be $K \leq \min(m, N)$. We consider the SVD of the transposed data matrix, $\mathbf{Z}_{N \times m} = \mathbf{Y}^T$, since in the PCA literature mentioned herein it is common to assume that the rows correspond to samples, and the columns correspond to the input variables. Note that $\mathbf{Z}_{N \times m}$ denotes an $N \times m$ matrix, and $\mathbf{I}_k$ denotes the $k \times k$ identity matrix. The SVD of $\mathbf{Z}$ is written as

$$\mathbf{Z} = \mathbf{U}\mathbf{D}\mathbf{A}^T, \quad \mathbf{U}^T\mathbf{U} = \mathbf{I}_N, \quad \mathbf{A}^T\mathbf{A} = \mathbf{I}_m, \quad d_1 \geq d_2 \geq ... \geq d_K > 0.$$

The well-known property of the SVD is that its first $k \leq K$ components (the first k columns of $\mathbf{U}$) produce the best approximation, in the sense of the Frobenius norm, to the matrix $\mathbf{Z}$, i.e.,

$$\sum_{i=1}^{k} d_i \mathbf{u}_i \mathbf{a}_i^T = \arg \min_{\hat{\mathbf{Z}} \in M(k)} ||\mathbf{Z} - \hat{\mathbf{Z}}||_2^2,$$

where $M(k)$ is the set of all $N \times m$ matrices of rank k.

The two views of PCA outlined above, the analysis (solving a sequence of eigen-value problems) and the synthesis (SVD, or matrix factorization), are equivalent, i.e., they find the same set of principal components. However, the equivalence does not hold anymore once additional constraints, such as sparsity, are added to the problem.

9.2.2   Sparse PCA: Synthesis View

Incorporating sparsity into PCA became a popular research direction, motivated by the goal of improving interpretability of the classical PCA approaches. Indeed, though PCA can reduce dimensionality of the data, capturing the data variability by a few components, the mapping from the input to the principal space still uses all the input variables, i.e., all loadings are nonzero. This reduces interpretability of the re-sults, especially if one attempts to identify the input variables that are most relevant. Several recently proposed sparse PCA approaches impose sparsity-enforcing con-straints on the loadings, thus achieving variable selection in the input space. We will first consider the synthesis-view, or matrix-factorization, approaches to sparse PCA as stated in eq. 9.3 above. Note that this sparse PCA formulation is closely related to the dictionary-learning (sparse coding) problem discussed above, with the main difference that the sparsity is enforced not on the code (components) matrix $\mathbf{X}$, but rather on the dictionary (loadings) matrix $\mathbf{A}$, as shown in Figure 9.5, as compared to the sparse-coding scheme shown in Figure 9.1.

The matrix-factorization formulation of sparse PCA was considered in several recent papers (Zou et al., 2006; Bach et al., 2008; Witten et al., 2009). A natural extension of the matrix-factorization formulation of PCA in order to enforce sparsity on the component loadings is to add the l_1-norm regularization on the columns of

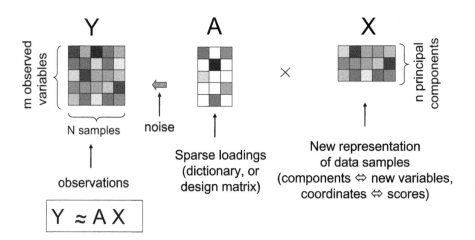

FIGURE 9.5: Sparse PCA as matrix factorization; note that the loadings (dictionary) matrix $\mathbf{A}$ is assumed to be sparse, as opposed to the code (components) matrix $\mathbf{X}$ in dictionary learning.

A, which leads to the following problem, considered, for example, by (Bach et al., 2008) and (Witten et al., 2009):

$$\min_{\mathbf{A},\mathbf{X}} ||\mathbf{Y} - \mathbf{A}\mathbf{X}||_2^2 + \lambda \sum_{i=1}^{k} ||\mathbf{a}_i||_1, \text{ subject to } ||\mathbf{x}_i||_2 \leq 1, \qquad (9.4)$$

where $\mathbf{a}_i$ and $\mathbf{x}_i$ denote the i-th column of $\mathbf{A}$ and i-th row of $\mathbf{X}$, respectively, and the dimensions of the matrices $\mathbf{A}$ and $\mathbf{X}$ are $m \times k$ and $k \times N$, respectively, k being the number of components. The problem is convex in $\mathbf{A}$ given $\mathbf{X}$, and vice versa, but non-convex in both $(\mathbf{A}, \mathbf{X})$, and thus alternating-minimization approaches are proposed; efficient methods based on simple column updates are considered, for example, in (Lee et al., 2006a; Witten et al., 2009).

Herein, we discuss in more detail the synthesis-view sparse PCA method of (Zou et al., 2006) that uses a connection between PCA and linear regression, and introduces the l_1-norm constraint to sparsify the loadings. More specifically, let $\mathbf{Y}^T = \mathbf{U}\mathbf{D}\mathbf{A}^T$ be the singular value decomposition (SVD) of the (transposed) data matrix $\mathbf{Y}^T$; then the rows of $\mathbf{X} = (\mathbf{U}\mathbf{D})^T$ are the principal components and the columns of $\mathbf{A}$ are the corresponding loadings. As discussed above, each principal component $\mathbf{x}_i$ is a linear combination of the m input variables, $\mathbf{x}_i = \mathbf{Y}^T\mathbf{a}_i$; thus, the loadings $\mathbf{a}_i$ can be found by regressing this component on the variables, as discussed, for example, in (Cadima and Jolliffe, 1995). In (Zou et al., 2006), this regression approach is extended to handle the $m > N$, using ridge regularizer; namely, given some positive λ, (Zou et al., 2006) show that the ridge regression estimate

$$\hat{\mathbf{w}} = \arg \min_{\mathbf{w}} ||\mathbf{x}_i - \mathbf{Y}^T\mathbf{w}||_2^2 + \lambda||\mathbf{w}||_2^2$$

recovers, after normalization, to the loadings of the i-th component, i.e. $\mathbf{a}_i = \hat{\mathbf{w}}/||\mathbf{w}||_2$. Then (Zou et al., 2006) enforce the sparsity on the loadings by adding the l_1-norm regularizer to the above ridge problem, and obtaining the Elastic Net regression formulation (Zou and Hastie, 2005). However, the above formulation cannot be used directly to find the loadings since the components $\mathbf{x}_i$ are not known. Instead, (Zou et al., 2006) derive the following criterion, that lends itself easily to an alternating minimization scheme presented in Figure 9.6. Let k denote the number of principal components, let $\mathbf{y}_i$ denote the i-th sample (column of $\mathbf{Y}$), $i = 1, ..., N$, $\mathbf{W}_{m \times k}$ denote an $m \times k$ matrix $\mathbf{W}$, and let $\mathbf{I}_k$ denote the $k \times k$ identity matrix. The following result holds:

Theorem 9.1. *(Zou et al., 2006) Let* $\mathbf{V}_{m \times k} = (\mathbf{v}_1, ...\mathbf{v}_k)$ *and* $\mathbf{W}_{m \times k} = (\mathbf{w}_1, ..., \mathbf{w}_k)$. *For any* $\lambda > 0$, *let* $\hat{\mathbf{V}}$ *and* $\hat{\mathbf{W}}$ *be the solutions of the following problem:*

$$(\hat{\mathbf{V}}, \hat{\mathbf{W}}) = \arg \min_{\mathbf{V},\mathbf{W}} \sum_{i=1}^{N} ||\mathbf{y}_i - \mathbf{W}\mathbf{V}^T\mathbf{y}_i||_2^2 + \lambda \sum_{i=1}^{k} ||\mathbf{w}_i||_2^2, \qquad (9.5)$$

$$\textit{subject to } \mathbf{V}^T\mathbf{V} = \mathbf{I}_k. \qquad (9.6)$$

Then $\hat{\mathbf{w}}_i$ *is proportional to the i-th component's loadings vector,* $\mathbf{a}_i$, *for* $i = 1, ..., k$.

Sparse PCA

Input: $m \times N$ matrix of samples $\mathbf{Y}$, number of principal components k, sparsity parameters λ, γ_i, for $i = 1, ...k$.

Initialize: Set $\mathbf{V} = \mathbf{A}_{PCA(k)}$, where the columns of $\mathbf{A}_{PCA(k)}$ are the loadings of the first k principal components obtained by the ordinary PCA, i.e., using SVD decomposition $\mathbf{Y}^T = \mathbf{UDA}^T$.

Alternating-minimization loop:

1. Given $\mathbf{V}_{m \times k} = (\mathbf{v}_1, ...\mathbf{v}_k)$, solve the Elastic Net for each $i = 1, ..., k$:

$$\mathbf{w}_i = \arg\min_{\mathbf{w}}(\mathbf{v}_i - \mathbf{w})^T \mathbf{YY}^T(\mathbf{v}_i - \mathbf{w}) + \lambda||\mathbf{w}||_2^2 + \gamma_i||\mathbf{w}||_1.$$

2. Given $\mathbf{W} = (\mathbf{w}_1, ..., \mathbf{w}_k)$, compute the SVD of $\mathbf{YY}^T\mathbf{W} = \mathbf{UDA}^T$.

3. Update: $\mathbf{V} = \mathbf{UA}^T$.

4. Repeat steps 1-3 until convergence.

5. Normalize the loadings: $\mathbf{a}_i = \mathbf{w}_i/||\mathbf{w}_i||_2^2$, for $i = 1, ..., k$, and return $\mathbf{A}$.

FIGURE 9.6: Sparse PCA algorithm based on the Elastic Net (Zou et al., 2006).

Given the above regression-like criterion, the sparsity on the loadings $\mathbf{a}_i$ is enforced by simply adding the l_1-norm regularizer to the above formulation, obtaining the following sparse PCA (SPCA) criterion (Zou et al., 2006):

$$(\hat{\mathbf{V}}, \hat{\mathbf{W}}) = \arg\min_{\mathbf{V},\mathbf{W}} \sum_{i=1}^{N} ||\mathbf{y}_i - \mathbf{WV}^T\mathbf{y}_i||_2^2 + \lambda \sum_{i=1}^{k} ||\mathbf{w}_j||_2^2 + \sum_{i=1}^{k} \gamma_j||\mathbf{w}_j||_1, \quad (9.7)$$

$$\text{subject to } \mathbf{V}^T\mathbf{V} = \mathbf{I}_k. \quad (9.8)$$

Note that using potentially different regularization parameters γ_i allows for different levels of sparsity in the loadings of different principal components. As it is common in matrix-factorization problems, the above criterion can be minimized using an alternating-minimization approach, as shown in Figure 9.6. However, since the objective function is not convex, we can only hope to find a good local minimum, as it is typically the case in matrix-factorization settings. Also, as noted by (Zou et al., 2006), the algorithm is not very sensitive to the choice of λ; in the case of $m < N$, no regularization is necessary and λ can be set to zero.

9.2.3 Sparse PCA: Analysis View

The second class of sparse PCA methods (see, for example, (Jolliffe et al., 2003; d'Aspremont et al., 2007, 2008)) follows the analysis view of PCA, where the ultimate objective is to find

$$\max_{\mathbf{a}} \mathbf{a}^T \mathbf{C} \mathbf{a} \text{ subject to } ||a||_2 = 1, ||a||_0 \leq k, \tag{9.9}$$

where $\mathbf{C} = \mathbf{Y}\mathbf{Y}^T$ is the empirical covariance matrix given the dataset $\mathbf{Y}$, as discussed above. The above problem is NP-hard due to the cardinality constraint, as shown in (Moghaddam et al., 2006) via reduction of the original NP-hard sparse regression (subset-selection with least-squares loss) to sparse PCA. The work by (Jolliffe et al., 2003) introduced an algorithm called SCoTLASS, that, similarly to Lasso, replaces l_0-norm by l_1-norm in the above formulation; however, the resulting optimization problem is still not convex and the method is computationally expensive. This motivated further work by (d'Aspremont et al., 2007), where a convex (semidefinite) relaxation to the above problem was proposed. Specifically, the problem in eq. 9.9 is first relaxed into

$$\max_{\mathbf{a}} \mathbf{a}^T \mathbf{C} \mathbf{a} \text{ subject to } ||\mathbf{a}||_2 = 1, ||\mathbf{a}||_1 \leq k^{1/2}, \tag{9.10}$$

and then into the following semidefinite program (SDP)

$$\max_{\mathbf{M}} tr(\mathbf{C}\mathbf{M}) \text{ subject to } tr(\mathbf{M}) = 1, \mathbf{1}^T |\mathbf{M}| \mathbf{1} \leq k, \mathbf{M} \succeq 0, \tag{9.11}$$

where $\mathbf{M} = \mathbf{a}\mathbf{a}^T$, $tr(\mathbf{M})$ denotes the trace of $\mathbf{M}$, and $\mathbf{1}$ denotes the vector of all ones. The above problem can be solved using, for example, Nesterov's smooth minimization approach (Nesterov, 2005), as discussed in (d'Aspremont et al., 2007), which yields the computational complexity $O(m^4\sqrt{\log m}/\epsilon)$, where m is the dimensionality of the input (the number of rows in $\mathbf{Y}$) and ϵ is the desired solution accuracy. Also, recent work by (d'Aspremont et al., 2008) proposes a more refined formulation of the semidefinite relaxation to the sparse PCA problem, and derives tractable sufficient conditions for optimality (e.g., testing global optimality of a given vector a using such conditions has $O(m^3)$ complexity); moreover, (d'Aspremont et al., 2008) derive a greedy algorithm that computes a full set of good approximate solutions for all sparsity levels (i.e., the number of nonzeros), with total complexity $O(m^3)$.

9.3 Sparse NMF for Blind Source Separation

We now give an example of a different approach to sparse matrix factorization – the sparse nonnegative matrix factorization (NMF) algorithm proposed by (Hoyer, 2004), which uses alternative constraints to enforce sparsity. NMF has a long history of applications in a wide range of fields, from chemometrics (Lawton and Sylvestre, 1971) and especially computer vision (see (Shashua and Hazan, 2005; Li et al., 2001; Guillamet and Vitrià, 2002; Ho, 2008) and references therein), to natural language processing (Xu et al., 2003; Gaussier and Goutte, 2005) and bioinformatics (Kim and

Park, 2007), to name a few applications. This approach was also successfully applied in the context of *blind source separation* problems in various signal-processing applications, as well as in less traditional problems such as computer network diagnosis discussed below.

The blind source separation (BSS) problem, as the name suggests, aims at reconstructing a set of unobserved signals (sources), represented by rows of a matrix $\mathbf{X}$, given the observed linear mixtures of these signals (rows of a matrix $\mathbf{Y}$), where the mixing matrix $\mathbf{A}$ is also unknown (thus, the source separation is "blind"). In other words, we have a matrix-factorization problem, where $\mathbf{Y}$ is approximated by $\mathbf{AX}$. An example of BSS is the famous "cocktail party" problem, where n speakers (sources) are present in the room with m microphones, and the task is to reconstruct what each of the speakers is saying, and how close he or she is to each of the microphones. The rows in the $m \times N$ matrix $\mathbf{Y}$ correspond to the microphones (samples/measurements), while the N columns correspond to dimensionality of the signals in their time domain, i.e., the number of samples. The rows of the $n \times N$ matrix $\mathbf{X}$ correspond to signals (individual speakers) that are linearly combined into microphone measurements via the $m \times n$ mixing matrix $\mathbf{A}$, where each entry $\mathbf{a}_{ij}$ corresponds to the distance of j-th speaker to the i-th microphone.

The BSS framework also found applications in other domains, such as, for example, performance monitoring and diagnosis in distributed computer networks and systems. Given the heterogeneous, decentralized, and often noncooperative nature of today's large-scale networks, it is impractical to assume that all statistics related to an individual system's components such as links, routers, or application-layer components can be collected for monitoring purposes. On the other hand, end-to-end measurements, such as test transactions, or probes (e.g., ping, traceroute, and so on), are typically cheap and easy to obtain. This realization gave rise to the field of *network tomography* (Vardi, 1996), which focuses on inference-based approaches to estimate unavailable network characteristics from available measurements.

In particular, (Chandalia and Rish, 2007) proposed a BSS-based approach to simultaneous discovery of network performance bottlenecks, such as network link delays, and the routing matrix (also called "dependency matrix" in a more general setting when the possible causes of performance degradations can include other elements of a distributed computer system, such as specific software components) from a collection of end-to-end probe results, using an analogy with the BSS problem. In this context, the signal matrix $\mathbf{X}$ represents unobserved delays at a potentially very large number of the network components, such as network links, the mixing matrix $\mathbf{A}$ corresponds to the unknown routing matrix, and each row of observations in $\mathbf{Y}$ corresponds to the time it takes a particular end-to-end test transaction to complete, which can be approximated by the sum of times spent at each of the components along the routing path, plus some unexpected noise, i.e. $\mathbf{Y} \approx \mathbf{AX}$.

In order to find the matrices A and X, one must to solve a (constrained) optimization problem that minimizes the reconstruction error between $\mathbf{Y}$ and $\hat{\mathbf{Y}}$, where $\hat{\mathbf{Y}} = \mathbf{AX}$, subject to some constraints on those matrices that are imposed by the application domain. There are several choices of loss functions to minimize the error, for example, the squared error or the KL-divergence. The constraints of nonnegativity on $\mathbf{A}$ and $\mathbf{X}$, as well as sparsity on the $\mathbf{X}$ columns, appear naturally, since both the routing matrix entries and the link delays are obviously nonnegative, and it is typically the case that only a very small number of network components are the performance bottlenecks. This combination of constraints yields the so-called sparse nonnegative matrix factorization (NMF) problem, often considered in the literature. Particularly, (Chandalia and Rish, 2007) evaluated two sparse NMF approaches, proposed by Hoyer (2004) and by Cichocki et al. (2006), respectively, in the context of the computer network data analysis.

Similarly to Hoyer (2004), the sparse NMF problem is formulated as

$$
\begin{aligned}
\min_{\mathbf{A},\mathbf{X}} \quad & \|\mathbf{Y} - \mathbf{A}\mathbf{X}\|_2^2 \\
\text{subject to} \quad & \text{sparsity}(\mathbf{a}_i) = s_A \forall i \in \{1, ..., m\} \\
& \text{sparsity}(\mathbf{x}_j) = s_X \forall j \in \{1, ..., N\},
\end{aligned}
\tag{9.12}
$$

where $\mathbf{a}_i$ is the i-th row of $\mathbf{A}$ and $\mathbf{x}_j$ is the j-th column of $\mathbf{X}$. The sparsity($\mathbf{u}$) of a d-dimensional vector $\mathbf{u}$ is defined as follows:

$$
\text{sparsity}(\mathbf{u}) = \frac{1}{\sqrt{d}-1}\left(\sqrt{d} - \frac{\sum |u_i|}{\sqrt{\sum u_i^2}}\right).
\tag{9.13}
$$

The desired sparsity levels s_A and s_X are given as an input. The above notion of sparsity varies smoothly between 0 (indicating minimum sparsity) and 1 (indicating maximum sparsity). It exploits the relation between l_1- and l_2-norms, thus giving great flexibility to achieve desired sparse solutions. The algorithm by (Hoyer, 2004) uses projected gradient descent, where in each iteration, matrices $\mathbf{A}$ and $\mathbf{X}$ are first updated by taking a step in the direction of the negative gradient, and then each row-vector of $\mathbf{A}$ and each column-vector of $\mathbf{X}$ are (non-linearly) projected onto a nonnegative vector with desired sparsity.

Empirical results on both simulated and real network topologies and network traffic, presented in (Chandalia and Rish, 2007), demonstrated that both sparse NMF methods were able to accurately reconstruct the routing matrix and the bottleneck locations, provided that the level of noise in the system was not too high. Figure 9.7, reproduced here from (Chandalia and Rish, 2007), plots the reconstruction accuracy for both the signal/link delay matrix $\mathbf{X}$ (panel (a)) and the routing matrix (panel (b)), in a simulation experiment, where the network traffic was generated by a simulator, but the network topology was real – a subnetwork of *Gnutella*, a large peer-to-peer network commonly used as a benchmark for experiments in the network tomography field. The "ground-truth" routing matrix $\mathbf{A}$ had 127 columns (nodes) and 50 rows (end-to-end probes). Different evaluation criteria were used for $\mathbf{A}$ and $\mathbf{X}$. Since $\mathbf{A}$ was subsequently binarized and interpreted as a routing matrix, the natural measure of accuracy was the average number of mistakes (0/1 flips) made by the reconstruction method as compared to the ground-truth routing matrix. On the other hand, the matrix $\mathbf{X}$ represented real-valued delays at different network nodes, and thus averaged correlation between the actual and reconstructed vectors of delays at particular nodes was used as a success measure; see (Chandalia and Rish, 2007) for

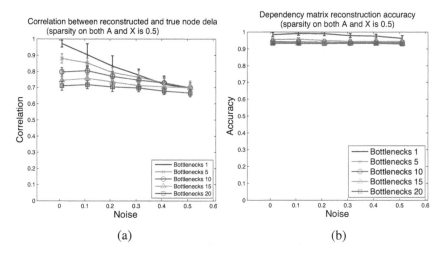

(a)   (b)

FIGURE 9.7: Recovering (a) performance bottlenecks and (b) routing table, or the so-called "dependency matrix" via sparse NMF applied to simulated traffic on Gnutella network.

more details on the experimental setup and definition of performance measures, including handling the permutations of rows and columns in the reconstructed matrices to match the ground truth. Overall, sample results in Figure 9.7 demonstrate that both $\mathbf{A}$ and $\mathbf{X}$ were reconstructed with a very high accuracy.

9.4   Summary and Bibliographical Notes

In this chapter, we departed from the standard sparse signal recovery setting where the design matrix (i.e., the dictionary) is given in advance. Instead, we considered the sparse matrix-factorization setting, which includes such popular problems as dictionary learning, or sparse coding (Olshausen and Field, 1996), as well as sparse PCA, where both the dictionary (loadings) and the signals, or code (components) must be learned, provided a set of data samples (measurements). In other words, the key change in the problem setting here is the inclusion of a set of hidden (unobserved) variables in the model; these hidden variables correspond to dictionary elements, and/or principle components.

Sparse coding, or dictionary learning, was originally introduced in the context of neuroscience (Olshausen and Field, 1996, 1997), and more recently became an active area of research in signal processing, statistics, and machine learning; some representative examples include the work by (Lewicki and Olshausen, 1999; Lewicki and Sejnowski, 2000; Engan et al., 1999; Kreutz-Delgado et al., 2003; Lesage et al., 2005; Elad and Aharon, 2006; Aharon et al., 2006a,b; Yaghoobi et al., 2009; Skretting and

Engan, 2010; Mairal et al., 2009, 2010; Tosic and Frossard, 2011; Lee et al., 2006a; Gregor and LeCun, 2010), and multiple other authors. Another example of sparse matrix factorization is sparse PCA, which includes several lines of work, based on either synthesis, or sparse matrix factorization, view of PCA (see, e.g., (Zou et al., 2006; Bach et al., 2008; Witten et al., 2009)), or analysis view, which involves solving a sequence of sparse eigenvalue problems (see, for example, (Jolliffe et al., 2003; d'Aspremont et al., 2007, 2008)). Yet another area involving sparse matrix factorization is the blind source separation in signal processing, where the goal is to simultaneously recover a set of sources/signals and the mixing matrix. We considered a particular type of approach to this problem, such as sparse nonnegative matrix factorization (NMF) of (Cichocki et al., 2006) and (Hoyer, 2004), and their somewhat non-traditional application to the performance bottleneck discovery in large-scale distributed computer systems and networks (Chandalia and Rish, 2007). Note that, in general, NMF is widely used in numerous applications including, but not limited to, chemometrics (Lawton and Sylvestre, 1971), computer vision (Shashua and Hazan, 2005; Li et al., 2001; Guillamet and Vitrià, 2002; Ho, 2008), natural language processing (Xu et al., 2003; Gaussier and Goutte, 2005), and bioinformatics (Kim and Park, 2007).

Finally, there are multiple recent developments related to sparse matrix factorization and its applications, including efficient optimization approaches to sparse coding (Lee et al., 2006a; Gregor and LeCun, 2010; Mairal et al., 2010), as well as extensions to more sophisticated types of sparse coding problems, such as: group sparse coding (Bengio et al., 2009; Garrigues and Olshausen, 2010), hierarchically structured dictionaries (Xiang et al., 2011), and other types of structured sparse coding (Szlam et al., 2011); Bayesian formulations with alternative sparsity priors, such as spike-and-slab (Shelton et al., 2012), or other, smoother priors such as KL-regularization (Bradley and Bagnell, 2008); non-parametric Bayesian approaches (Zhou et al., 2009), nonlinear sparse coding (Shelton et al., 2012; Ho et al., 2013), and smooth sparse coding via kernel smoothing and marginal regression (Balasubramanian et al., 2013), to name a few. Recent applications of sparse coding range from visual recognition problems (Morioka and Shiníchi, 2011), music signal representation (Dikmen and Févotte, 2011), and spatio-temporal feature learning in videos, such as motion capture data (Kim et al., 2010), to novel document detection in online data streams (Kasiviswanathan et al., 2012), as well as multitask and transfer learning (Maurer et al., 2013), among many other applications.

Epilogue

Sparse modeling is a rapidly growing area of research that emerged at the intersection of statistics and signal processing. The popularity of sparse modeling can be attributed to the fact that a seemingly impossible task of reconstructing a high-dimensional unobserved signal from a relatively small number of measurements becomes feasible when a particular type of signal structure is present. This structure, referred to as sparsity, assumes that most of the signal's dimensions are zero or close to zero, in some basis. Surprisingly, this assumption holds for many natural signals. Moreover, efficient algorithms based on a convex relaxation of the NP-hard sparse recovery problem are available, making sparse modeling practical in many real-life applications.

As we already mentioned, no single book cannot capture all recent developments in this vast and constantly expanding field. Herein, we attempted to cover some of its key theoretical and algorithmic aspects, and provided various application examples, with a particular focus on statistical analysis of neuroimaging. Bibliographical sections at the end of each chapter include references to various sparsity-related topics that we did not discuss in full detail.

We hope that our book will serve as a good introduction to the exciting new field of sparse modeling, which, at the same time, is deeply rooted in the ancient principle of parsimony, and can even be related to our daily lives. You have probably heard this popular inspirational story: a professor shows an empty jar to his students, and fills it with a few large rocks; he then dumps some gravel in to fill the empty spaces, and, finally, pours in sand, filling the jar completely. The point is clear: remember to focus on a few most-important "big rocks" in your life, before so many other, much less important "gravel" and "sand" variables fill up your time.

Appendix

Mathematical Background

A.1   Norms, Matrices, and Eigenvalues

Herein, we will use the notation $\mathbb{Z}_N = \{0, \ldots, n-1\}$ to index the coordinates of an n-dimensional vector. We will also use $\mathbf{A}^T$ and $\mathbf{x}^T$ to denote the transpose of a matrix $\mathbf{A}$ and of a vector $\mathbf{x}$, respectively. $\mathbf{A}^*$ will denote the *conjugate transpose*, or *adjoint matrix*, of a matrix $\mathbf{A}$ with complex entries; recall that the conjugate transpose of $\mathbf{A}$ is obtained from $\mathbf{A}$ by taking the transpose and then taking the complex conjugate of each entry (i.e., negating their imaginary parts but not their real parts).

We will denote by $< \mathbf{x}, \mathbf{y} >$ the *inner product* of the two complex vectors $\mathbf{x} \in \mathbb{C}^N$ and $\mathbf{y} \in \mathbb{C}^N$. When the vectors are real-valued, it is equivalent to the dot product (or scalar product) $\sum_{i \in \mathbb{Z}_N} x_i y_i$, and can be also written as a matrix product $\mathbf{x}^T \mathbf{y}$, assuming that both $\mathbf{x}$ and $\mathbf{y}$ are column-vectors.

Let $\mathbf{x} \in \mathbb{C}^N$. The l_q-*norm* of $\mathbf{x}$, denoted $||\mathbf{x}||_q$, is defined for $q \geq 1$ as (also, see chapter 2):

$$||\mathbf{x}||_q = \left(\sum_{n \in \mathbb{Z}_N} |x_n|^q \right)^{1/q}. \tag{A.1}$$

It is easy to verify that for $q \geq 1$, the function $||\mathbf{x}||_q$ defined above is indeed a proper norm, i.e., it satisfies the norm properties, such as:

1. zero vector norm: $||\mathbf{x}|| = 0$ if and only if $\mathbf{x} = 0$;

2. absolute homogeneity: $\forall \alpha, \alpha \neq 0, ||\alpha \mathbf{x}|| = |\alpha| ||\mathbf{x}||$; and

3. triangle inequality: $||\mathbf{x} + \mathbf{y}|| \leq ||\mathbf{x}|| + ||\mathbf{y}||$.

When $0 < q < 1$, the function defined in eq. A.1 is not a proper norm since it violates the triangle inequality. Indeed, let $\mathbf{x} = (1, 0, ..., 0)$ and $\mathbf{y} = (0, 1, 0, ..., 0)$ be two unit vectors in R^n. Then, for any $0 < q < 1$, we have $||\mathbf{x}||_q + ||\mathbf{y}||_q = 2$, but $||\mathbf{x} + \mathbf{y}||_q = 2^{1/q} > 2$, i.e. $||\mathbf{x} + \mathbf{y}||_q > ||\mathbf{x}||_q + ||\mathbf{y}||_q$, which violates the triangle inequality. However, for convenience sake, even when $0 < q < 1$, the function $||\mathbf{x}||_q$ is still frequently called the l_q-norm, despite some abuse of terminology. For the case of $q = 0$, we denote by $||\mathbf{x}||_0$ the size of the *support* of $\mathbf{x}$, denoted $supp(\mathbf{x})$, defined as a set of nonzero coordinates of $\mathbf{x}$. Thus, $||\mathbf{x}||_0 = |supp(\mathbf{x})|$.

We now define some matrix properties. Let $\mathbf{A}$ be an $N \times M$ matrix with real or complex entries. When $M = N$, i.e., when the matrix is square, the *trace* of a matrix is defined as the sum of its diagonal elements:

$$Tr(\mathbf{A}) = \sum_{i \in \mathbb{Z}_N} a_{ii}.$$

Note that

$$Tr(\mathbf{AB}) = Tr(\mathbf{BA}). \tag{A.2}$$

Next, we define the $||.||_q$ *norm of a matrix* as

$$||\mathbf{A}||_q = \sup_{\mathbf{x}, ||\mathbf{x}||_q = 1} ||\mathbf{Ax}||_q.$$

In particular,

$$||\mathbf{A}||_1 = \max_{j \in \mathbb{Z}_M} \sum_{i \in \mathbb{Z}_N} |a_{ij}|, \; and \; ||\mathbf{A}||_\infty = \max_{i \in \mathbb{Z}_N} \sum_{j \in \mathbb{Z}_M} |a_{ij}|. \tag{A.3}$$

For $1 \leq p \leq \infty$, $||\mathbf{A}||_q$ is a norm. Another quite useful matrix norm is the Perron-Frobenius norm defined as

$$||\mathbf{A}||_F = \sqrt{\sum_{i \in \mathbb{Z}_N j \in \mathbb{Z}_M} |a_{ij}|^2} = \sqrt{Tr(\mathbf{A}^* \mathbf{A})}, \tag{A.4}$$

where $\mathbf{A}^*$ is the conjugate transpose of $\mathbf{A}$.

Note that concepts and definitions summarized herein can be also found in any standard linear algebra course; however, they are included here for completeness sake.

A.1.1   Short Summary of Eigentheory

Given an $N \times N$ matrix $\mathbf{A}$, an *eigenvector* $\mathbf{x}$ of $\mathbf{A}$ is a vector satisfying $\mathbf{Ax} = \lambda\mathbf{x}$, for some complex number λ that is called an *eigenvalue* of $\mathbf{A}$. The set of all eigenvalues of matrix $\mathbf{A}$, denoted by $Sp(\mathbf{A})$, is called the *spectrum of A*. The *kernel* of $\mathbf{A}$, also called its *null space* or *nullspace*, denoted by $Ker(\mathbf{A})$ or $N(\mathbf{A})$, is the set of all vectors $\mathbf{x}$, called *null vectors*, that satisfy $\mathbf{Ax} = 0$. Matrix $\mathbf{A}$ is called *degenerate*, or *singular*, if and only if its kernel includes a nonzero vector (i.e., there exists $\mathbf{x} \in Ker(\mathbf{A})$ such that $\mathbf{x} \neq 0$), or, equivalently, if and only if

$$0 \in Sp(\mathbf{A}). \tag{A.5}$$

For example, any nonzero null vector of $\mathbf{A}$ (i.e., a nonzero vector from the kernel of $\mathbf{A}$) is an eigenvector with zero eigenvalue.

By the *Cayley-Hamilton theorem*, each eigenvalue is a root of the *characteristic equation* $det(\mathbf{A} - \lambda\mathbb{I}) = 0$, where $\mathbb{I}$ is the *identity matrix* of size $N \times N$, i.e. the matrix with ones on the main diagonal and zeros elsewhere. There is at least one eigenvector for each root of the characteristic equation. For a *diagonal square matrix*, i.e. the matrix with nonzero main diagonal and zeros elsewhere, every element on the diagonal is an eigenvalue. The *multiplicity* of an eigenvalue λ_i as a root of the characteristic equation, i.e., the largest integer k such that $(\lambda - \lambda_i)^k$ divides that polynomial, is called the *algebraic multiplicity* of λ_i, denoted m_{λ_i}. For each eigenvalue λ_i, there are exactly m_{λ_i} linearly independent vectors corresponding to that eigenvalue, i.e. vectors x satisfying $(\mathbf{A} - \lambda_i\mathbb{I})^{m_{\lambda_i}} x = 0$. Note that not all of those vectors are eigenvectors.

If a square $N \times N$ matrix $\mathbf{A}$ is self-adjoint (i.e., $\mathbf{A}^* = \mathbf{A}$), then $det(\mathbf{A} - \lambda\mathbb{I}) = det(\mathbf{A}^* - \bar{\lambda}\mathbb{I})$; hence, every eigenvalue is real, and there are exactly N eigenvectors. In this case, two eigenvectors $\mathbf{u}_i$ and $\mathbf{u}_j$ corresponding to different eigenvalues λ_i and λ_j will be orthogonal, since

$$\lambda_i < \mathbf{u}_i, \mathbf{u}_j > = < \mathbf{A}\mathbf{u}_i, \mathbf{u}_j > = < \mathbf{u}_i, \mathbf{A}\mathbf{u}_j > = \lambda_j < \mathbf{u}_i, \mathbf{u}_j >,$$

which holds only when $< \mathbf{u}_i, \mathbf{u}_j > = 0$, i.e. for orthogonal $\mathbf{u}_i$ and $\mathbf{u}_j$. We can think of a square matrix $\mathbf{A}$ as a set of row-vectors in a basis defined by the row-vectors $e_1 = (1, 0, ..., 0); e_2 = (0, 1, 0, ..., 0); e_N = (0, ..., 0, 1)$; similarly, the set of column-vectors of $\mathbf{A}$ corresponds to the basis of $e_1^T, ..., e_N^T$. If the basis is changed, then the matrix $\mathbf{A}$ is transformed into $\mathbf{P}\mathbf{A}\mathbf{P}^{-1}$, where $\mathbf{P}$ is an invertible matrix (*linear operator*) that corresponds to the basis change.

If an $N \times N$ matrix $\mathbf{A}$ has exactly N independent eigenvectors, it becomes a diagonal matrix, denoted $\mathbf{D}$, when written in the basis of the eigenvectors. Indeed, for each eigenvalue λ_i and the associated eigenvector $\mathbf{u}_i$ from the basis, we have

$$\mathbf{A}\mathbf{u}_i = \lambda_i\mathbf{u}_i. \tag{A.6}$$

In other words, multiplying a basis vector $\mathbf{u}_i$ by the matrix $\mathbf{A}$ is equivalent to multiplying $\mathbf{u}_i$ by a scalar λ_i, and hence $\mathbf{D}$ ($\mathbf{A}$ transformed into the basis of eigenvectors) is a diagonal matrix with diagonal entries equal to the corresponding eigenvalues.

Let $\mathbf{A}$ be a square self-adjoint matrix. By combining the facts stated above with respect to eigenvectors with the property of the trace given by eq. A.2, we conclude that the trace of a square self-adjoint matrix $\mathbf{A}$ is simply the sum of $\mathbf{A}$'s eigenvalues, i.e.,

$$Tr(\mathbf{A}) = Tr(\mathbf{P}\mathbf{D}\mathbf{P}^{-1}) = Tr(\mathbf{P}^{-1}\mathbf{P}\mathbf{D}) = Tr(\mathbf{D}) = \sum_{i \in \mathbb{Z}_N} \lambda_i. \tag{A.7}$$

A.2 Discrete Fourier Transform

In this section we present the Discrete Fourier Transform (DFT), a discrete transform widely used in signal processing and related fields; essentially, DFT converts a set of (equally spaced) function samples from the original domain, such as time, into different domain, namely, the frequency domain.

There are two variants of DFT, the complex DFT and the real DFT (RDFT or CT, see below). We focus primarily on the complex DFT, and only mention here the definition of the real DFT.

Definition 11. *The DFT is defined on a finite sequence (a vector) of N real or complex numbers* $\mathbf{x} = (x_0, x_1, \ldots, x_{N-1})'$ *as a sequence (a vector)* $\mathbf{X} = \mathcal{F}(\mathbf{x})$ *with the coordinates given by*

$$X_k = \sum_{n=0}^{N-1} x_n e^{-2\pi i \frac{k}{N} n}.$$

The transformation $\mathcal{F}$ is a linear transformation on $\mathbb{C}^N$ defined by a matrix with entries $\mathcal{F}_{n,k} = e^{-2\pi \frac{k}{N} n}$. The columns $\mathbf{u}_i$ of the matrix $\mathcal{F}$ are mutually orthogonal, since

$$< \mathbf{u}_i, \mathbf{u}_j > = \sum_{n=0}^{N-1} e^{2\pi \frac{i}{N} n} e^{-2\pi i \frac{j}{N} n} = \sum_{n=0}^{N-1} e^{2\pi i \frac{i-j}{N} n} = N\delta_{i,j}, \qquad \text{(A.8)}$$

where $\delta_{i,j}$ denotes the Kronecker delta function. Recall that the form $<,>$ is conjugate linear, i.e., a semi-linear form by the first argument.

Equation A.8 implies that the inverse $\mathcal{F}^{-1}$ of the function $\mathcal{F}$ is proportional to its conjugate $\frac{1}{N}\mathcal{F}^*$. It also leads to *Plancherel's identity*:

$$< \mathbf{x}, \mathbf{y} > = \frac{1}{N} < \mathbf{X}, \mathbf{Y} > . \qquad \text{(A.9)}$$

Indeed,

$$\begin{aligned} < \mathbf{x}, \mathbf{y} > = \ &< \mathbf{x}, \mathcal{F}^{-1}\mathcal{F}(\mathbf{y}) > = < \mathbf{x}, \tfrac{1}{N}\mathcal{F}^*\mathcal{F}(\mathbf{y}) > \\ = \ &\tfrac{1}{N} < \mathcal{F}(\mathbf{x}), \mathcal{F}(\mathbf{y}) > = \tfrac{1}{N} < \mathbf{X}, \mathbf{Y} > . \end{aligned} \qquad \text{(A.10)}$$

This immediately implies *Parseval's identity*, by setting $\mathbf{x} = \mathbf{y}$:

$$< \mathbf{x}, \mathbf{x} > = \frac{1}{N} < \mathbf{X}, \mathbf{X} > . \qquad \text{(A.11)}$$

In order to keep real-valued the result of applying DFT to real-valued vectors, one can just use the real (or just the imaginary) part of the DFT. Namely:

Definition 12. *The Real DFT (RDFT) is defined on a finite sequence (a vector) of N real numbers* $\mathbf{x} = (x_0, x_1, \ldots, x_{N-1})'$ *as a sequence* $\mathbf{X}$ *with the coordinates*

$$X_k = \sum_{n=0}^{N-1} x_n \cos(\frac{2\pi kn}{N}), \ or \ X_k = \sum_{n=0}^{N-1} x_n \sin(\frac{2\pi kn}{N}).$$

The first transform is called the Discrete Cosine Transform (DCT) and the second one the Discrete Sine Transform (DST). Herein, we will mainly use DCT.

Without going into further details, we just would like to note that analogs of the equalities A.8, A.9, and A.11 can be also established for for DCT and DST.

A.2.1 The Discrete Whittaker-Nyquist-Kotelnikov-Shannon Sampling Theorem

The sampling theorem, associated with the names of Whittaker, Nyquist, Kotelnikov, and Shannon, gives criteria for a continuous signal to have a discrete finite spectrum and to be exactly reconstructable from its discrete finite spectrum. While the original Shannon proof is sufficiently transparent, a rigorous proof requires elaborate results from the Fourier transform theory. Here we concentrate on the discrete version of the theorem.

Given a discrete signal of size N, $\mathbf{x} = (x_n | n \in \mathbb{Z}_N)$, we assume that the observable part of the spectrum or DFT of the signal has the support K_1, $\mathbf{X}|_{K_1} = \mathcal{F}(\mathbf{x})|_{K_1}$. We also assume that any reconstruction tool (decoder) can be used. Our question is: what are the conditions on K_1 so that the original signal $\mathbf{x}$ can be reconstructed exactly?

Theorem A.1. *((Discrete-to-discrete) Whittaker-Nyquist-Kotelnikov-Shannon sampling theorem) Let $\mathbf{x} = (x_n | n \in \mathbb{Z}_N)$ be an N-dimensional signal, and let K_1 be the support of the observable part of the spectrum $\mathbf{X} = \mathcal{F}(x)$. Then $\mathbf{x}$ can be recovered exactly if and only if $|K_1| = N$.*

Proof. If $|K_1| = N$, then by using as a decoder D an inverse DFT, i.e., $D = \mathcal{F}^{-1}$, the signal $\mathbf{x}$ is recovered exactly: $\mathbf{x} = \mathcal{F}^{-1}\mathcal{F}(\mathbf{x})$. If $|K_1| < N$, then there is a one-point part of the spectrum, say, $n_0 \in \mathbb{Z}_N$, that is not observable. Consider all N-dimensional signals with the spectrum in the n_0, $\mathcal{L} = \{\mathbf{y} = (y_n | n \mathbb{Z}_N, supp(\mathcal{F}(\mathbf{x})) = n_0\}$. The set $\mathcal{L}$ is not empty since $\mathcal{L} = \mathcal{F}^{-1}((0, \ldots, 0, \lambda, 0, \ldots, 0)$, where λ is a real number at the n_0-th position. Any two signals that differ on an element from $\mathcal{L}$ will be mapped by the DFT $\mathcal{F}$ to the same spectrum observable on K_1, and thus could not be decoded differently. $\square$

Finally, we also state the well-known discrete-to-continuous version of the above sampling theorem:

Theorem A.2. *((Discrete-to-continuous) Whittaker-Nyquist-Kotelnikov-Shannon sampling theorem) A uniformly sampled analog signal can be recovered perfectly as long as the sampling rate is at least twice as large as the highest-frequency component of the signal.*

A.3   Complexity of l_0-norm Minimization

In (Natarajan, 1995) it was shown that the l_0-norm minimization problem is *NP-hard* (see (Garey and Johnson, 1979) for the definition and examples of NP-hard problems). Herein, we also provide a brief proof sketch for this important fact. We formulate below the Minimum Relevant Variables problem (which is essentially the l_0-norm minimization problem) and the 3-set cover problem (called MP5 and X3C in (Garey and Johnson, 1979), respectively), and show the direct reduction of X3C to MP5. Note that this consideration may be also extended to the approximation results. See, for example, (Ausiello et al., 1999), page 448, or (Schrijver, 1986).

Definition 13. *(Minimum Relevant Variables, MP5) Instance: A matrix* **A** *of size* $n \times m$ *with integer coefficients and a vector* **b** *of size* n *with integer coefficients.*

Question: What is the minimal number of nonzeros that an n*-dimensional vector* **x** *with rational coefficients can have, given that* **x** *is the solution of* $\mathbf{Ax} = \mathbf{b}$*?*

Definition 14. *(Exact Cover by 3 sets, X3C) Instance: A finite set* X *with* $|X| = 3q$ *and a collection* C *of 3-element subsets of* X*.*

Question: Does C *contain an exact cover for* X*, or a sub-collection* $C' \subseteq C$ *such that every element of* X *occurs in exactly one element of* C'*?*

The problem X3C is one of the basic problems in the set of NP-complete problems stated in (Garey and Johnson, 1979). The restriction in the MP5 problem for the matrix **A** and the vector **b** to have integer coefficients is not essential; indeed, since MP5 is concerned with finding *rational* coefficients of **x**, we can multiply all entries of **A** and **b** by the common denominator (of all entries of **A** and **b**). In order to show that, MP5 problem is NP-complete, we map the problem X3C into the problem MP3 ((Garey and Johnson, 1979) refer to this part as to an unpublished result). We give the details of the mapping here for completeness sake. The set X with size $|X| = 3q$ we map into $\mathbb{R}^{3q}$, subsets S_i from S we map into matrix $\mathbf{A} = (a_{j,i})$ of size $3q \times |S|$, where $a_{j,i} = 1$ if and only if S_i covers point corresponding to the j-th coordinate of $\mathbb{R}^{3q}$, and 0 otherwise. Vector **x** is of size $|S|$ and corresponds to the choosen subsets of S: ($x_j = 1$ if and only if the covering subset S_j is choosen, and 0 otherwise. As a vector **b** we choose the vector with all $3q$ coefficients equal to 1. Assuming this mapping, the solution of the X3C becomes the minimal solution of MP5, since the number of nonzero coefficients should be at least q.

A.4   Subgaussian Random Variables

Subgaussian random variables are important because they generalize nice properties of finitely supported random variables and those of Gaussian random variables to a much wider class of random variables.

Definition 15. *Let X be a real random variable with $\mathrm{E}X = 0$. We say that X has a-subgaussian tail if there exists a constant $a, C > 0$ such that for all $\lambda > 0$,*

$$Prob(|X| > \lambda) \le Ce^{-a\lambda^2}. \tag{A.12}$$

We say that X has a-subgaussian tail up to λ_0 if the previous bound holds for all $\lambda \le \lambda_0$.

If X_1, X_2, ..., X_n is a sequence of random variables, by saying that they have a uniform subgaussian tail we mean that all of them have subgaussian tails with the same constant a.

We shorten the term X *having a-subgaussian tail* to *subgaussian random variable X.*

Example 3. *Let X be a random variable with finite support. An example of such random variable is the Bernoulli random variable. Then for $t > M = max|supp(|X|)|$, we have $Prob(|X| > M) = 0$, and thus it is a subgaussian variable. Another example of a subgaussian X is simply a Gaussian random variable.*

The (moment) generating function of a random variable X is a random variable Y defined as $Y = e^{uX}$, with $Prob(Y = e^{uX} > e^{ut}) = Prob(X > t)$. The generating function is an important tool in the probability theory. The generating function of the normal (Gaussian) distribution $\mathcal{N}(\mu, \sigma)$ is $e^{\mu t + \frac{1}{2}\sigma^2 t^2}$. We characterize the subgaussian random variable using its generating function.

Proposition A.3. *(Characterization of a normalized subgaussian random variable) Let X be a random variable with $\mathrm{E}X = 0$ and $\mathrm{E}X^2 < \infty$.*

i) *If $\mathrm{E}e^{u|X|} \le Ce^{Cu^2}$ for some constant $C > 0$ and for all $u > 0$, then X has a subgaussian tail. If $\mathrm{E}e^{u|X|} \le Ce^{Cu^2}$ holds for all $u \in (0, u_0]$, then X has a subgaussian tail up to $2Cu_0$.*

ii) *If X has a subgaussian tail, then $\mathrm{E}e^{u|X|} \le e^{Cu^2}$ for all $u > 0$, with the constant C depending only on the constant a of the subgaussian tail.*

Proof. i). For all $u \in (0, u_0]$ and all $t \ge 0$, we have by the Markov inequality $Prob(X > t) \le \frac{\mathrm{E}(X)}{t}$:

$$Prob(|X| \ge t) = Prob(e^{u|X|} \ge e^{ut}) \tag{A.13}$$

$$\text{(Markov inequality)} \quad \le e^{-ut}\mathrm{E}e^{u|X|} \tag{A.14}$$

$$\le Ce^{-ut+Cu^2}. \tag{A.15}$$

For $t \le 2Cu_0$, setting $u = t/2C$ in the above estimate implies $Prob[X \ge t] \le Ce^{-t^2/4C}$.

ii). Let F be the distribution function of X; in other words, $F(t) = Prob(X < t)$.

We have $Ee^{u|X|} = \int_{-\infty}^{\infty} e^{u|t|} dF(t)$. We split the integration interval into two subintervals, corresponding to $ut \leq 1$ and $ut \geq 1$. Then

$$\int_{-1/u}^{1/u} e^{u|t|} dF(t) \leq \int_{-1/u}^{1/u} 3 dF(t) \leq 3. \qquad (A.16)$$

We estimate the second integral by sum for both positive and negative parts:

$$\int_{1/u}^{\infty} e^{u|t|} dF(t) \leq \sum_{k=1}^{\infty} e^{k+1} Prob(X \geq \frac{k}{u})$$

$$\leq C \sum_{k=1}^{\infty} e^{2k} e^{-ak^2/u^2} = C \sum_{k=1}^{\infty} e^{k(2-ak/u^2)}, \qquad (A.17)$$

$$\int_{-\infty}^{-1/u} e^{u|t|} dF(t) \leq \sum_{k=1}^{\infty} e^{k+1} Prob(X \leq -\frac{k}{u})$$

$$C \leq \sum_{k=1}^{\infty} e^{2k} e^{-ak^2/u^2} = C \sum_{k=1}^{\infty} e^{k(2-ak/u^2)}. \qquad (A.18)$$

For $u \leq \sqrt{a}/2$, we have $2 - ak/u^2 \leq -a/2u^2$ and the sum is bounded by the geometric series with both first term and quotient $e^{-a/2u^2} \leq e^{-1} < \frac{1}{2}$. So the sum is at most $2e^{-a/2u^2} = O(u^2)$ (since $e^x \geq 1 + x > x$, take the reciprocal values to obtain $e^{-x} \leq \frac{1}{x}$ for $x > 0$, and substitute $x = a/2u^2$). Hence $Ee^{u|X|} \leq 1 + O(u^2) \leq Ce^{O(u^2)}$.

For $u > \sqrt{a}/2$, the largest terms in the considered sum are those with k near u^2/a, and the sum is $O(e^{u^2/2a})$. So $Ee^{u|X|} \leq Ce^{O(u^2)}$ holds. □

Example 4. *For the random variable X with finite support $|X| \leq M$ for some positive M, we have an estimate*

$$Ee^{uX} \leq Ee^{uM} = e^{uM}. \qquad (A.19)$$

Hence for $M' = \max\{(\frac{M}{2})^2, e\}$ holds

$$ln M' + M'u^2 \geq 2\sqrt{ln M'\, M'u^2} \geq 2u\sqrt{M'} \geq Mu, \qquad (A.20)$$

or

$$Ee^{uX} \leq M' e^{M'u^2}. \qquad (A.21)$$

In other words, random variable X satisfies Proposition A.3 ii) with constant $C = max\{(\frac{M}{2})^2, e\}$.

Theorem A.4. *Let $X_0, ..., X_{n-1}$ be a set of the subgaussian random variables with constant C as in Theorem A.3 ii). Then*

$$E(\max_{i \in \mathbb{Z}_n} |X_i|) \leq 2\sqrt{C \log(2n)} \qquad (A.22)$$

Proof. For each $u > 0$, the following holds:

$$exp(u \, E(\max_{i \in \mathbb{Z}_n} |X_i|)) \leq E \max_{i \in \mathbb{Z}_n} e^{u \, |X_i|}$$

$$\leq E(\sum_{i \in \mathbb{Z}_n} e^{u \, X_i} + e^{-u \, X_i}) \leq 2n e^{Cu^2}. \tag{A.23}$$

By setting $u = \sqrt{\frac{log2n}{C}}$ we get A.22. $\qquad\square$

Corollary A.5. *Let X_i be random variables with finite support $|X_i| \leq M, M > e$ and 0 mean. Then*

$$E(\max_{i \in \mathbb{Z}_n} |X_i|) \leq M. \tag{A.24}$$

The Gaussian random variables are 2-stable, i.e. the mixture of two normal variables (with 0 mean and standard deviation $\sigma = 1$), where the squared mixture coefficients sum to 1, is again normal (see the proposition below). Our goal is to establish that subgaussian random variables with 0 mean and $\sigma = 1$ are closed with respect to such a mixture.

Corollary A.6. *(Mixture of subgaussian random variables) Let X_1, ..., X_n be independent random variables, satisfying $\mathrm{E}X_i = 0$, Var $X_i < C < \infty$, and having a uniform subgaussian tail. Let α_1, ..., α_n be real coefficients satisfying $\alpha_1^2 + \cdots + \alpha_n^2 = 1$. Then the sum*

$$Y = \alpha_1 X_1 + \cdots + \alpha_n X_n \tag{A.25}$$

has $\mathrm{E}Y = 0$, Var $Y < \infty$, and a-subgaussian tail.

Proof. We have $\mathrm{E}Y = 0$ by linearity of expectation, and since the variance is additive for independent random variables,

$$VarY = \sum_{i=1}^{n} \alpha_i^2 VarX_i = \sum_{i=1}^{n} \alpha_i^2 < \infty. \tag{A.26}$$

Since $\mathrm{E}e^{uX_i} \leq Ce^{Cu^2}$ by Proposition A.3, we have

$$\mathrm{E}e^{uY} = \prod_{i=1}^{n} \mathrm{E}e^{u\alpha_i X_i} \leq Ce^{Cu^2(\alpha_1^2+\cdots+\alpha_n^2)} = Ce^{Cu^2}, \tag{A.27}$$

and hence Y has a subgaussian tail. $\qquad\square$

Sometimes we are interested in a linear combination of subgaussian random variables. The following corollary states that a linear combination of the zero-mean subgaussian random variables is also subgaussian.

Corollary A.7. *(Azuma/Hoefding inequality) Let X_1, ..., X_n be independent random variables, satisfying $\mathrm{E}X_i = 0$, Var $X_i < \infty$, and having a uniform subgaussian tail. Let $\alpha = \{\alpha_1, ..., \alpha_n\}$ be real coefficients. Then the sum $\sum \alpha_i X_i$ satisfies*

$$Prob(|\sum \alpha_i X_i| > t) \leq Ce^{-\frac{Ct^2}{||\alpha||_{l_2}^2}}. \tag{A.28}$$

Proof. Indeed,

$$Prob(|\sum \alpha_i X_i| > t) \quad = Prob(|\sum \tfrac{\alpha_i}{||\alpha||_{l_2}} X_i| > \tfrac{t}{||\alpha||_{l_2}}) \tag{A.29}$$

$$\leq Ce^{-\tfrac{Ct^2}{||\alpha||_{l_2}^2}}, \tag{A.30}$$

where the last inequality follows from Proposition A.6.   □

Proposition A.8. *Let $k \geq 1$ be an integer. Let $Y_1, \ldots, Y_k$ be independent random variables with $EY_i = 0$, Var $Y_i = 1$, and a uniform subgaussian tail. Then $Z = \frac{1}{\sqrt{k}}(Y_1^2 + Y_2^2 + \cdots + Y_k^2 - k)$ has a subgaussian tail up to $\sqrt{k}$.*

To establish A.8 we first show the following:

Lemma A.9. *If Y is as the Y_i in Proposition A.8, then there are constants C and u_0 such that for all $u \in [0, u_0]$ we have $Ee^{u(Y^2-1)} \leq e^{Cu^2}$ and $Ee^{u(1-Y^2)} \leq e^{Cu^2}$.*

We begin with the first inequality. Note that EY^4 is finite (a constant); this follows from the subgaussian tail of Y, estimate $t^4 = O(e^t + e^{-t})$ for all t, and Proposition A.3, ii).

Let F be the distribution function of Y^2; that is, $F(t) = \text{Prob}\ (Y^2 < t)$. Split the integral defining Ee^{uY^2} into two intervals, corresponding to $uY^2 \leq 1$ and $uY^2 \geq 1$. Thus,

$$Ee^{uY^2} = \int_0^{1/u} e^{ut} dF(t) + \int_{1/u}^{\infty} e^{ut} dF(t).$$

The first integral is estimated by

$$\int_0^{1/u} 1 + ut + u^2 t^2 dF(t) \leq \int_0^{\infty} 1 + ut + u^2 t^2 dF(t)$$

$$= 1 + uEY^2 + u^2 EY^4 = 1 + u + O(u^2).$$

The second integral can be estimated by a sum:

$$\sum_{k=1}^{\infty} e^{k+1}\, \text{Prob}\ (Y^2 \geq k/u) \leq 2 \sum_{k=1}^{\infty} e^{2k} e^{-ak/u}.$$

Assume that $u \leq u_0 = a/4$; then $k(2 - a/u) \leq -ka/2u$, and the sum is of order $e^{-\Omega(1/u)}$. Similar to the proof of Proposition A.3 we can bound this by $O(u^2)$, and for Ee^{uY^2} hence the estimate $1 + u + O(u^2) \leq e^{u+O(u^2)}$.

Then $Ee^{u(Y^2-1)} = Ee^{uY^2} e^{-u} \leq e^{O(u^2)}$.

The calculation for estimating Ee^{-uY^2} is simpler, since $e^{-ut} \leq 1 - ut + u^2 t^2$ for all $t > 0$ and $u > 0$:

$$\mathrm{E}e^{-uY^2} = \int_0^\infty e^{-ut} \mathrm{d}F(t) \le \int_0^\infty 1 - ut + u^2 t^2 \mathrm{d}F(t)$$

$$= 1 - u\mathrm{E}[Y^2] + u^2 \mathrm{E}Y^4 \le 1 - u + O(u^2) \le e^{-u+O(u^2)}.$$

This yields $\mathrm{E}e^{u(1-Y^2)} \le e^{O(u^2)}$.

Proof of Proposition A.8. For $Z = \dfrac{1}{\sqrt{k}}(Y_1^2 + \cdots + Y_k^2 - k)$ and $0 < u \le u_0\sqrt{k}$, with u_0 as in Lemma A.9, we calculate $\mathrm{E}[e^{uZ}] = \mathrm{E}[e^{(u/\sqrt{k})(Y_1^2+\cdots+Y_k^2-k)}] = \mathrm{E}[e^{(u/\sqrt{k})(Y^2-1)}]^k \le (e^{Cu^2/k})^k = e^{Cu^2}$. The proposition A.3 implies that Z has a subgaussian upper tail up to $2C\sqrt{k} \ge \sqrt{k}$ (assuming that $2C \ge 1$). The calculation for the lower tail is similar.

A.5 Random Variables and Symmetrization in $\mathbb{R}^n$

We consider now the symmetrization of random variables in $\mathbb{R}^n$. This type of argument is used frequently to reduce an arbitrary random variable to the random variable symmetrized with the Bernoulli random variable taking values $\{\pm 1\}$ with equal probability.

A random vector $X = (x_j)_{j\in\mathbb{Z}_d}$ in $\mathbb{R}^n$ is a measurable mapping from event (probabilistic) space (Ω, μ) into the $\mathbb{R}^n$; here μ is some probabilistic measure on $\mathbb{R}^n$. The distribution function of X is $D((a_j)_{j\in\mathbb{Z}_d}) = \mu(\{\omega \in \Omega | X(\omega) \le (a_j)_{j\in\mathbb{Z}_d}\})$ for the point $a = (a_j)_{j\in\mathbb{Z}_d} \in \mathbb{R}^d$. The expectation EX is defined as

$$EX = \int_\Omega X(\omega)d\mu(\omega) \in \mathbb{R}^d, \tag{A.31}$$

where the integral is calculated coordinate-wise. The notion of variance is replaced with *covariance* defined as

$$\Sigma X = EXX' = EX \otimes X, \tag{A.32}$$

with Σ being semi-definite $d \times d$ matrix.

Frequently, as an events space (Ω, μ) the space $\mathbb{R}^d$ with measure defined by distribution function $D((a_j)_{j\in\mathbb{Z}_d})$ is choosen.

Definition 16. *The vectors $X_i, i \in \mathbb{Z}_N$ are independent if for any parallelepiped $I_i = \prod_{j\in\mathbb{Z}_d}[a'_{ji}, a''_{ji}]$ with $a'_{ji} \le a''_{ji}$ holds*

$$\mu(\{\omega | X_i(\omega) \in I_i \text{ for each } i \in \mathbb{Z}_N\}) = \prod_{i\in\mathbb{Z}_N} \mu(\{\omega | X_i(\omega) \in I_i\}). \tag{A.33}$$

Since the definition of the independence uses only an image of X_i, the independence of $X_i, i \in \mathbb{Z}_N$ implies the independence of $f(X_i), i \in \mathbb{Z}_N$ for any function $f : \mathbb{R}^d \mapsto \mathbb{R}^p$.

The following statement may be found in (Ledoux and Talagrand, 2011; Vershynin, 2012).

Lemma A.10. *Let $X_i, i \in \mathbb{Z}_N$ be a finite set of independent random variables with values in $\mathbb{R}^d$ with some norm $||||$. Then*

$$E|| \sum_i (X_i - EX_i)|| \leq 2E|| \sum_i \epsilon_i X_i ||. \tag{A.34}$$

If X, Y are independent identically distributed random variables with values in $\mathbb{R}^d$ with some norm $|| \cdot ||$, then

$$E||X - EX|| \leq E||X - Y|| \leq 2\, E||X - EX||, \tag{A.35}$$

and for all positive u holds

$$Prob(||X|| > 2EX + u) \leq 2\, Prob(||X - Y|| > u). \tag{A.36}$$

Proof. Since the variables X_i are independent, the realization probability space for the variables $\epsilon_i X_i$ will be product $\prod_{i\in\mathbb{Z}_N}(\Omega, \mu) \times \prod_{i\in\mathbb{Z}_N}(B, \nu)$ of n copies of (Ω, μ) with n copies of space (B, ν), where $B = \{-1, 1\}, \mu = (1/2, 1/2)$.

We also consider variables Y_i being independent copies of variable X_i. The common realization space will be $\prod_{i\in\mathbb{Z}_N}(\Omega, \mu) \times \prod_{i\in\mathbb{Z}_N}(B, \nu) \times \prod_{i\in\mathbb{Z}_N}(\Omega, \mu)$ with variables X_i dependent only on the first n coordinates, variables ϵ_i dependent only on the second n coordinates, and Y_i dependent only on the last coordinates. We denote expectations by the first(second/third) n coordinates by E_X, E_ϵ, E_Y accordingly. Then

$$E|| \sum_{i\in\mathbb{Z}_N}(X_i - EX_i)|| = E|| \sum_{i\in\mathbb{Z}_N}(X_i - EY_i)||$$

$$\leq E_X E_Y || \sum_{i\in\mathbb{Z}_N}(X_i - Y_i)|| = E_X E_Y || \sum_{i\in\mathbb{Z}_N} \epsilon_i(X_i - Y_i)||$$

$$= E_\epsilon E_X E_Y || \sum_{i\in\mathbb{Z}_N} \epsilon_i(X_i - Y_i)||$$

$$\leq E_X E_Y E_\epsilon || \sum_{i\in\mathbb{Z}_N} \epsilon_i(X_i)|| + E_X E_Y E_\epsilon || \sum_{i\in\mathbb{Z}_N} \epsilon_i(Y_i)||$$

$$= 2E|| \sum_{i\in\mathbb{Z}_N} \epsilon_i X_i ||,$$

where the first equality is due to the fact that variables X_i and Y_i are identically distributed and independent.

The first inequality is the corollary of the triangle inequality and is known as the Jensen inequality. It states that for two vectors, the norm of half sum does not exceed half sum of norm, and then it is extended to the expectation of norm of random variables vs. norm of expectation of vector.

The second equality is due to the fact that by multiplying $X_i - Y_i$ by -1 we simply exchange the i-th variable in the first and the third n-tuple of omegas, and, hence, the total expectation remains the same.

The third equality is valid since we find an expectation of a constant function. The second inequality is due to the triangle inequality for norm, and the possibility to change the order of integration due to the Fubini theorem. Finally, the last equality is valid since $\epsilon_i X_i$ and $\epsilon_i Y_i$ are identically distributed.

To show A.35, we first use three expressions in the proof of A.34 and then just use triangle inequality:

$$
\begin{aligned}
E||X - EX|| &= E||(X - EY)|| \le E_X E_Y ||X - Y|| = E||X - Y|| \\
&\le E||X - EX|| + E||Y - EY|| = 2E||X - EX||.
\end{aligned}
\tag{A.37}
$$

To show A.36, we use independence of X and Y and the fact that $Prob(||X|| \ge 2E||X||) \, 2E||X|| \le E||X||$, and hence $Prob(||X|| \le 2E||X||) \ge 1/2$:

$$
\begin{aligned}
Prob(||X - Y|| > u) &\ge Prob(||X|| - ||Y|| > u) \\
&\ge Prob(\{||X|| > 2E||X|| + u\} \cap \{||Y|| \le 2E||Y||\}) \\
&= Prob(||Y|| \le 2E||Y||) Prob(||X|| > 2E||X|| + u) \\
&\ge 1/2 \, Prob(||X|| > 2E||X|| + u).
\end{aligned}
\tag{A.38}
$$

$\square$

A.6 Subgaussian Processes

Let (T, d) be a metric space, in other words, T is a set and d is a distance on T.

Definition 17. *Subgaussian process $V_t, t \in T$ with indices in the metric space (T, d) is a set of random variables, possibly with values in $\mathbb{R}^q$ satisfying for some constant c and each $t, t' \in T$*

$$
Prob(\{||V_t - V_t'|| > s\}) \le e^{\frac{cs^2}{d^2(t,t')}}.
\tag{A.39}
$$

Example 5. *Let $\psi_i, i \in \mathbb{Z}_q$ be a set of independent subgaussian random variables; choose $(T, d) = (\mathbb{R}^q, ||\,||_{\mathbb{R}^q})$. Then by Azuma/Hoefding inequality $V_t, t \in T$:*

$$
V_t = \sum_{i \in \mathbb{Z}_q} t_i \psi_i
\tag{A.40}
$$

is a subgaussian process.

A.7 Dudley Entropy Inequality

The goal of this section is to give proof of the Dudley inequality. We follow Talagrand (1996) and Rudelson (2007) in our exposition. Loosely speaking, the inequality estimates probability of divergence of random process through integral of metric entropy of index set. In order to introduce the inequality we need to clarify what $\sup_{t \in T} |V_t - V_{t_0}|$ means. Since each process is defined a.e. taking sup may end up not being a measurable function.

We assume that the index set T is compact, in order to have a uniformly bounded distance between any two sets of points $t, t' \in T$, and a bounded diameter of $d(T) = \sup_{t,t' \in T} d(t, t')$. Recall that being compact means that every cover by open balls contains finite sub-cover. The equivalent formulation says that every continuous function reaches its sup at some point.

There is a countable dense in (T, d) set, since by covering T with open balls of size 2^{-i}, and taking finite sub-cover, we obtain a dense, countable subset. We take supremum over that set. This extra assumption is sufficient to keep us in the measurable world.

Recall also that $log N(T, d, \epsilon)$ is called metric entropy of metric space (T, d).

Theorem A.11. *Let (T, d) be a compact metric space. Let $V_t, t \in T$ be a subgaussian process, and let $M = \int_0^\infty \sqrt{\log N(T, d, \epsilon)} d\epsilon$. Then for any $s \geq 1$, and any $t_0 \in T$,*

$$Prob(\{\sup_{t \in T} ||V_t - V_{t_0}|| > sM\}) \leq e^{-C \cdot s^2}. \tag{A.41}$$

Proof. By property of compact, T may be covered with a finite number of unit balls, hence there exists a natural number j_0 such that $d(T) \leq 2^{j_0}$. Consider

$$M_2 = \sum_{j=-j_0}^{\infty} 2^{-j} \sqrt{\log N(T, d, \epsilon)}. \tag{A.42}$$

We have

$$d(T) \leq 2^{-j_0} \leq M_2,$$

and, since $\sqrt{\log N(T, d, \epsilon)}$ is decreasing function by ϵ,

$$\frac{1}{2} M_2 \leq M \leq 2M_2. \tag{A.43}$$

Thus, up to change of constant C in A.41 it is enough to show that

$$Prob(\{\sup_{t \in T} ||V_t - V_{t_0}|| > sM_2\}) \leq e^{-C \cdot s^2}. \tag{A.44}$$

We choose Π_j being $2^{-j} - \epsilon$-net with N_j points and approximate t by $\pi_j(t) \in \Pi_j$, one of the closest points of Π_j to t. As a first point $\pi_{j_0}(t)$ we choose t_0.

We represent $V_t - V_{t_0}$ as the chain sum

$$V_t - V_{t_0} = V_t - V_{\pi_l(t)} + \sum_{j=j_0+1}^{l} (V_{\pi_j(t)} - V_{\pi_{j-1}(t)}), \qquad (A.45)$$

and estimate the contribution of each summand in A.45 using subgaussian tail inequality. Since $\pi_j(t)$ is the closest point to t in $2^{-j} - \epsilon$-net, then

$$d(\pi_k(t), \pi_{k-1}(t)) \leq d(\pi_k(t), t) + d(t, \pi_{k-1}(t)) \leq 2^{-k+2}.$$

There are at most $N_k N_{k-1} \leq N_k^2$ pairs $(\pi_k(t), \pi_{k-1}(t))$.

Let a_j be a sequence of positive numbers to be defined later. By definition of subgaussian tail, for any $x \in T$,

$$Prob(\{||V_{\pi_j(x)} - V_{\pi_{j-1}(x)}|| \geq a_j\}) \leq e^{\frac{ca_j^2}{16 \cdot 2^{2j}}}.$$

The probability of supremum in A.44 does not exceed

$$\sum_{j=j_0}^{\infty} N_j N_{j-1} e^{\frac{ca_j^2}{16 \cdot 2^{2j}}} \leq \sum_{j=j_0}^{\infty} N_j^2 e \cdot \frac{ca_j^2}{16 \cdot 2^{2j}}. \qquad (A.46)$$

Choose $a_j = \frac{4}{\sqrt{c}} 2^{-j} \cdot (\sqrt{logN_j} + \sqrt{j - j_0 + 2} + s)$. Then the right side of A.46 does not exceed

$$\sum_{j=j_0}^{\infty} e^{-(j-j_0+2)-s^2} \leq e^{-s^2}.$$

We established that

$$Prob(\{\sup_{t \in T} ||V_t - v_{t_0}|| \geq \sum_{j=j_0}^{\infty} a_j\}) \leq e^{-s^2}.$$

Since

$$\sum_{j=j_0}^{\infty} a_j \leq C \sum_{j=j_0}^{\infty} 2^{-j}(\sqrt{logN_j} + s\sqrt{j - j_0 + 2}) \leq CM_2 + sd(T) + CM_2 \leq CM_2 \cdot s,$$

we completed the proof of A.41. □

Corollary A.12. *(Dudley entropy inequality) Let $V_t, t \in T$ be a subgaussian process with $V_{t_0} = 0$. Then*

$$E \sup_{t \in T} ||V_t|| \leq C \int_0^{\infty} \sqrt{logN(T, d, \epsilon)} d\epsilon. \qquad (A.47)$$

Proof. Using Theorem A.11 we get

$$E \sup_{t \in T} ||V_t|| = \int_0^{\infty} xProb(\{\sup_{t \in T} ||V_t|| > x\})dx \leq CM + \int_{CM}^{\infty} xe^{-x^2/CM} dx \leq CM, \qquad (A.48)$$

for appropriate choice of constants on each step. □

A.8   Large Deviation for the Bounded Random Operators

We need the following large deviation type estimate for the uniformly bounded operators. See Theorems 6.17 and 6.19 from Ledoux and Talagrand (2011) for $s = Kl$.

Theorem A.13. *Let $Y_1, ..., Y_n$ be independent symmetric random variables in $\mathbb{R}^n$. Assume that $||Y_j|| \leq M$. Then for any $l \geq q$, and any $t > 0$, the random variable $Y = ||\sum_{j \in \mathbb{Z}_n} Y_j||$ satisfies*

$$Prob(Y \geq 8qE(Y) + 2Ml + t) \leq \frac{C}{q^l} + 2e^{-\frac{t^2}{256E(Y)^2}}. \qquad (A.49)$$

Bibliography

Adamczak, R., Litvak, A., Pajor, A., Tomczak-Jaegermann, N., 2011. Restricted isometry property of matrices with independent columns and neighborly polytopes by random sampling. Constructive Approximation 34 (1), 61–88.

Aharon, M., Elad, M., Bruckstein, A., 2006a. K-SVD: An algorithm for designing overcomplete dictionaries for sparse representation. IEEE Transactions on Signal Processing 54 (11), 4311–4322.

Aharon, M., Elad, M., Bruckstein, A., 2006b. On the uniqueness of overcomplete dictionaries, and a practical way to retrieve them. Linear algebra and its applications 416 (1), 48–67.

Antoniadis, A., Fan, J., 2001. Regularization of wavelet approximations. Journal of the American Statistical Association 96 (455).

Asadi, N. B., Rish, I., Scheinberg, K., Kanevsky, D., Ramabhadran, B., 2009. MAP approach to learning sparse Gaussian Markov networks. In: Proc. of the IEEE International Conference on Acoustics, Speech and Signal Processing (ICASSP). pp. 1721–1724.

Asif, M., Romberg, J., 2010. On the Lasso and Dantzig selector equivalence. In: Proc. of the 44th Annual Conference on Information Sciences and Systems (CISS). IEEE, pp. 1–6.

Atia, G., Saligrama, V., March 2012. Boolean compressed sensing and noisy group testing. IEEE Transactions on Information Theory 58 (3), 1880–1901.

Ausiello, G., Protasi, M., Marchetti-Spaccamela, A., Gambosi, G., Crescenzi, P., Kann, V., 1999. Complexity and Approximation: Combinatorial Optimization Problems and Their Approximability Properties. Springer-Verlag New York.

Bach, F., 2008a. Bolasso: Model consistent Lasso estimation through the bootstrap. In: Proc. of the 25th International Conference on Machine Learning (ICML). pp. 33–40.

Bach, F., 2008b. Consistency of the group Lasso and multiple kernel learning. Journal of Machine Learning Research 9, 1179–1225.

Bach, F., June 2008c. Consistency of trace norm minimization. Journal of Machine Learning Research 9, 1019–1048.

Bach, F., 2010. Self-concordant analysis for logistic regression. Electronic Journal of Statistics 4, 384–414.

Bach, F., Jenatton, R., Mairal, J., Obozinski, G., 2012. Optimization with sparsity-inducing penalties. Foundations and Trends in Machine Learning 4 (1), 1–106.

Bach, F., Lanckriet, G., Jordan, M., 2004. Multiple kernel learning, conic duality, and the SMO algorithm. In: Proc. of the Twenty-first International Conference on Machine Learning (ICML).

Bach, F., Mairal, J., Ponce, J., 2008. Convex sparse matrix factorizations. arXiv preprint arXiv:0812.1869.

Bakin, S., 1999. Adaptive regression and model selection in data mining problems. Ph.D. thesis, Australian National University, Canberra, Australia.

Balasubramanian, K., Yu, K., Lebanon, G., 2013. Smooth sparse coding via marginal regression for learning sparse representations. In: Proc. of the International Conference on Machine Learning (ICML). pp. 289–297.

Baliki, M., Geha, P., Apkarian, A., 2009. Parsing pain perception between nociceptive representation and magnitude estimation. Journal of Neurophysiology 101, 875–887.

Baliki, M., Geha, P., Apkarian, A., Chialvo, D., 2008. Beyond feeling: Chronic pain hurts the brain, disrupting the default-mode network dynamics. The Journal of Neuroscience 28 (6), 1398–1403.

Banerjee, A., Merugu, S., Dhillon, I., Ghosh, J., April 2004. Clustering with Bregman divergences. In: Proc. of the Fourth SIAM International Conference on Data Mining. pp. 234–245.

Banerjee, A., Merugu, S., Dhillon, I. S., Ghosh, J., October 2005. Clustering with Bregman divergences. Journal of Machine Learning Research 6, 1705–1749.

Banerjee, O., El Ghaoui, L., d'Aspremont, A., March 2008. Model selection through sparse maximum likelihood estimation for multivariate Gaussian or binary data. Journal of Machine Learning Research 9, 485–516.

Banerjee, O., Ghaoui, L. E., d'Aspremont, A., Natsoulis, G., 2006. Convex optimization techniques for fitting sparse Gaussian graphical models. In: Proc. of the 23rd International Conference on Machine Learning (ICML). pp. 89–96.

Baraniuk, R., Davenport, M., DeVore, R., Wakin, M., 2008. A simple proof of the restricted isometry property for random matrices. Constructive Approximation 28 (3), 253–263.

Beck, A., Teboulle, M., 2009. A fast iterative shrinkage-thresholding algorithm for linear inverse problems. SIAM J. Imaging Sciences 2 (1), 183–202.

Bengio, S., Pereira, F., Singer, Y., Strelow, D., 2009. Group sparse coding. In: Proc. of Neural Information Processing Systems (NIPS). Vol. 22. pp. 82–89.

Bertsekas, D., 1976. On the Goldstein-Levitin-Polyak gradient projection method. IEEE Transactions on Automatic Control 21 (2), 174–184.

Besag, J., 1974. Spatial interaction and the statistical analysis of lattice systems. Journal of the Royal Statistical Society. Series B (Methodological) 36 (2), 192–236.

Beygelzimer, A., Kephart, J., Rish, I., 2007. Evaluation of optimization methods for network bottleneck diagnosis. In: Proc. of the Fourth International Conference on Autonomic Computing (ICAC). Washington, DC, USA.

Beygelzimer, A., Rish, I., 2002. Inference complexity as a model-selection criterion for learning Bayesian networks. In: Proc. of the International Conference on Principles of Knowledge Representations and Reasoning (KR). pp. 558–567.

Bickel, P., December 2007. Discussion: The Dantzig selector: Statistical estimation when p is much larger than n. The Annals of Statistics 35 (6), 2352–2357.

Bickel, P., Ritov, Y., Tsybakov, A., 2009. Simultaneous analysis of Lasso and Dantzig selector. The Annals of Statistics 37 (4), 1705–1732.

Blomgren, P., Chan, F., 1998. Color TV: total variation methods for restoration of vector-valued images. IEEE Transactions on Image Processing 7 (3), 304–309.

Blumensath, T., Davies, M. E., 2007. On the difference between orthogonal matching pursuit and orthogonal least squares. Unpublished manuscript.

Borwein, J., Lewis, A., Borwein, J., Lewis, A., 2006. Convex analysis and nonlinear optimization: Theory and examples. Springer, New York.

Boyd, S., Vandenberghe, L., 2004. Convex Optimization. Cambridge University Press, New York, NY, USA.

Bradley, D., Bagnell, J., 2008. Differentiable sparse coding. In: Proc. of Neural Information Processing Systems (NIPS). pp. 113–120.

Buhl, S., 1993. On the existence of maximum likelihood estimators for graphical Gaussian models. Scandinavian Journal of Statistics 20 (3), 263–270.

Bühlmann, P., van de Geer, S., 2011. Statistics for High-Dimensional Data: Methods, Theory and Applications. Springer.

Bunea, F., 2008. Honest variable selection in linear and logistic regression models via l_1 and $l_1 + l_2$ penalization. Electron. J. Statist. 2, 1153–1194.

Bunea, F., Tsybakov, A., Wegkamp, M., 2007. Sparsity oracle inequalities for the Lasso. Electron. J. Statist. 1, 169–194.

Cadima, J., Jolliffe, I., 1995. Loading and correlations in the interpretation of principle compenents. Journal of Applied Statistics 22 (2), 203–214.

Cai, T., Liu, W., Luo, X., 2011. A constrained l_1 minimization approach to sparse precision matrix estimation. Journal of American Statistical Association 106, 594–607.

Cai, T., Lv, J., December 2007. Discussion: The Dantzig selector: Statistical estimation when p is much larger than n. The Annals of Statistics 35 (6), 2365–2369.

Candès, E., 2008. The restricted isometry property and its implications for compressed sensing. Comptes Rendus Mathematique 346 (9), 589–592.

Candès, E., Plan, Y., 2011. A probabilistic and RIPless theory of compressed sensing. IEEE Transactions on Information Theory 57 (11), 7235–7254.

Candès, E., Recht, B., 2009. Exact matrix completion via convex optimization. Foundations of Computational Mathematics 9 (6), 717–772.

Candès, E., Romberg, J., Tao, T., February 2006a. Robust uncertainty principles: Exact signal reconstruction from highly incomplete frequency information. IEEE Trans. on Information Theory 52 (2), 489–509.

Candès, E., Romberg, J., Tao, T., 2006b. Stable signal recovery from incomplete and inaccurate measurements. Communications on Pure and Applied Mathematics 59 (8), 1207–1223.

Candès, E., Tao, T., December 2005. Decoding by linear programming. IEEE Trans. on Information Theory 51 (12), 4203–4215.

Candès, E., Tao, T., 2006. Near optimal signal recovery from random projections: Universal encoding strategies? IEEE Trans. Inform. Theory 52 (12), 5406–5425.

Candès, E., Tao, T., 2007. The Dantzig selector: Statistical estimation when p is much larger than n. Annals of Statistics 35 (6), 2313–2351.

Carl, B., 1985. Inequalities of Bernstein-Jackson-type and the degree of compactness of operators in banach spaces. Annales de l'institut Fourier 35 (3), 79–118.

Carroll, M., Cecchi, G., Rish, I., Garg, R., Rao, A., 2009. Prediction and interpretation of distributed neural activity with sparse models. NeuroImage 44 (1), 112–122.

Cecchi, G., Huang, L., Hashmi, J., Baliki, M., Centeno, M., Rish, I., Apkarian, A., 2012. Predictive dynamics of human pain perception. PLoS Computational Biology 8 (10).

Cecchi, G., Rish, I., Thyreau, B., Thirion, B., Plaze, M., Paillere-Martinot, M.-L., Martelli, C., Martinot, J.-L., Poline, J.-B., 2009. Discriminative network models of schizophrenia. In: Proc. of Neural Information Processing Systems (NIPS). Vol. 22. pp. 250–262.

Chafai, D., Guédon, O., Lecué, G., Pajor, A., 2012. Interactions between compressed sensing, random matrices, and high dimensional geometry. forthcoming book.

Chan, T., Shen, J., 2005. Image Processing and Analysis: Variational, Pde, Wavelet, and Stochastic Methods. Society for Industrial and Applied Mathematics, Philadelphia, PA, USA.

Chandalia, G., Rish, I., 2007. Blind source separation approach to performance diagnosis and dependency discovery. In: Proc. of the 7th ACM SIGCOMM Conference on Internet Measurement (IMC). pp. 259–264.

Chen, S., Donoho, D., Saunders, M., 1998. Atomic decomposition by basis pursuit. SIAM Journal on Scientific Computing 20 (1), 33–61.

Cheraghchi, M., Guruswami, V., Velingker, A., 2013. Restricted isometry of Fourier matrices and list decodability of random linear codes. SIAM Journal on Computing 42 (5), 1888–1914.

Cichocki, A., Zdunek, R., Amari, S., 2006. New algorithms for non-negative matrix factorization in applications to blind source separation. In: Proc. of the IEEE International Conference on Acoustics, Speech and Signal Processing. Vol. 5. pp. 621–624.

Cohen, A., Dahmen, W., DeVore, R., 2009. Compressed sensing and best k-term approximation. Journal of the American Mathematical Society 22 (1), 211–231.

Collins, M., Dasgupta, S., Schapire, R., 2001. A generalization of principal component analysis to the exponential family. In: Proc. of Neural Information Processing Systems (NIPS). MIT Press.

Combettes, P., Pesquet, J.-C., 2011. Fixed-Point Algorithms for Inverse Problems in Science and Engineering. Chapter: Proximal Splitting Methods in Signal Processing. Springer-Verlag.

Combettes, P., Wajs, V., 2005. Signal recovery by proximal forward-backward splitting. SIAM Journal on Multiscale Modeling and Simulation 4, 1168–1200.

Cowell, R., Dawid, P., Lauritzen, S., Spiegelhalter, D., 1999. Probabilistic Networks and Expert Systems. Springer.

Cox, D., Wermuth, N., 1996. Multivariate Dependencies: Models, Analysis and Interpretation. Chapman and Hall.

Dai, W., Milenkovic, O., 2009. Subspace pursuit for compressive sensing reconstruction. IEEE Trans. Inform. Theory 55 (5), 2230–2249.

d'Aspremont, A., Bach, F. R., Ghaoui, L. E., 2008. Optimal solutions for sparse principal component analysis. Journal of Machine Learning Research 9, 1269–1294.

d'Aspremont, A., Ghaoui, L. E., Jordan, M. I., Lanckriet, G. R. G., 2007. A direct formulation for sparse PCA using semidefinite programming. SIAM Review 49 (3), 434–448.

Daubechies, I., Defrise, M., Mol, C. D., 2004. An iterative thresholding algorithm for linear inverse problems with a sparsity constraint. Communications on Pure and Applied Mathematics 57, 1413–1457.

Dempster, A. P., March 1972. Covariance selection. Biometrics 28 (1), 157–175.

Dikmen, O., Févotte, C., 2011. Nonnegative dictionary learning in the exponential noise model for adaptive music signal representation. In: Proc. of Neural Information Processing Systems (NIPS). pp. 2267–2275.

Do, T., Gan, L., Nguyen, N., Tran, T., 2008. Sparsity adaptive matching pursuit algorithm for practical compressed sensing. In: Proc. of the 42nd Asilomar Conference on Signals, Systems, and Computers. pp. 581–587.

Donoho, D., April 2006a. Compressed sensing. IEEE Trans. on Information Theory 52 (4), 1289–1306.

Donoho, D., July 2006b. For most large underdetermined systems of linear equations, the minimal l_1-norm near-solution approximates the sparsest near-solution. Communications on Pure and Applied Mathematics 59 (7), 907–934.

Donoho, D., June 2006c. For most large underdetermined systems of linear equations, the minimal l_1-norm solution is also the sparsest solution. Communications on Pure and Applied Mathematics 59 (6), 797–829.

Donoho, D., 2006d. For most large underdetermined systems of linear equations the minimal l_1-norm solution is also the sparsest solution. Comm. Pure Appl. Math. 59 (6), 797–829.

Donoho, D., Elad, M., 2003. Optimally sparse representation in general (nonorthogonal) dictionaries via l_1 minimization. Proc. Natl. Acad. Sci. USA 100, 2197–2202.

Donoho, D., Elad, M., Temlyakov, V., 2006. Stable recovery of sparse overcomplete representations in the presence of noise. IEEE Trans. Inform. Theory 52 (1), 6–18.

Donoho, D., Huo, X., 2001. Uncertainty principles and ideal atomic decomposition. IEEE Trans. Inform. Theory 47, 2845–2862.

Donoho, D., Stark, P., 1989. Uncertainty principles and signal recovery. SIAM J. Appl. Math. 49, 906–931.

Donoho, D., Tanner, J., 2009. Observed universality of phase transitions in high-dimensional geometry, with implications for modern data analysis and signal processing. Philosophical Transactions of the Royal Society A: Mathematical, Physical and Engineering Sciences 367 (1906), 4273–4293.

Donoho, D., Tsaig, Y., Drori, I., Starck, J., 2012. Sparse solution of underdetermined systems of linear equations by stagewise orthogonal matching pursuit. IEEE Transactions on Information Theory 58 (2), 1094–1121.

Dorfman, R., 1943. The detection of defective members of large populations. The Annals of Mathematical Statistics 14 (4), 436–440.

Du, D., Hwang, F., 2000. Combinatorial group testing and its applications, 2nd edition. World Scientific Publishing Co., Inc., River Edge, NJ.

Duchi, J., Gould, S., Koller, D., 2008. Projected subgradient methods for learning sparse Gaussians. In: Proc. of Uncertainty in Artificial Intelligence (UAI).

Edwards, D., 2000. Introduction to Graphical Modelling, 2nd Edition. Springer.

Efron, B., Hastie, T., 2004. LARS software for R and Splus: http://www.stanford.edu/~hastie/Papers/LARS/.

Efron, B., Hastie, T., Johnstone, I., Tibshirani, R., 2004. Least angle regression. Ann. Statist. 32 (1), 407–499.

Efron, B., Hastie, T., Tibshirani, R., December 2007. Discussion: The Dantzig selector: Statistical estimation when p is much larger than n. The Annals of Statistics 35 (6), 2358–2364.

Elad, M., 2006. Why simple shrinkage is still relevant for redundant representations? IEEE Transactions on Information Theory 52, 5559–5569.

Elad, M., 2010. Sparse and Redundant Representations: From Theory to Applications in Signal and Image Processing. Springer.

Elad, M., Aharon, M., 2006. Image denoising via sparse and redundant representations over learned dictionaries. IEEE Transactions on Image Processing 15 (12), 3736–3745.

Elad, M., Matalon, B., Zibulevsky, M., 2006. Image denoising with shrinkage and redundant representations. In: Proc. of the IEEE Computer Society Conference on Computer Vision and Pattern Recognition (CVPR). Vol. 2. IEEE, pp. 1924–1931.

Eldar, Y., Kutyniok, G. (editors), 2012. Compressed Sensing: Theory and Applications. Cambridge University Press.

Engan, K., Aase, S., Husoy, H., 1999. Method of optimal directions for frame design. In: Proc. of the International Conference on Acoustics, Speech, and Signal Processing (ICASSP). Vol. 5. pp. 2443–2446.

Fan, J., Li, R., 2005. Variable selection via nonconcave penalized likelihood and its oracle properties. Journal of the American Statistical Association 96, 1348–1360.

Fazel, M., Hindi, H., Boyd, S., 2001. A rank minimization heuristic with application to minimum order system approximation. In: Proc. of the 2001 American Control Conference. Vol. 6. IEEE, pp. 4734–4739.

Figueiredo, M., Nowak, R., 2003. An EM algorithm for wavelet-based image restoration. IEEE. Trans. Image Process. 12, 906–916.

Figueiredo, M., Nowak, R., 2005. A bound optimization approach to wavelet-based image deconvolution. In: Proc. of the IEEE International Conference on Image Processing (ICIP). Vol. 2. IEEE, pp. II–782.

Foucart, S., Pajor, A., Rauhut, H., Ullrich, T., 2010. The Gelfand widths of l_p-balls for $0 < p \leq 1$. Journal of Complexity 26 (6), 629–640.

Foucart, S., Rauhut, H., 2013. A mathematical introduction to compressive sensing. Springer.

Frank, I., Friedman, J., 1993. A statistical view of some chemometrics regression tools. Technometrics 35 (2), 109–148.

Friedlander, M., Saunders, M., December 2007. Discussion: The Dantzig selector: Statistical estimation when p is much larger than n. The Annals of Statistics 35 (6), 2385–2391.

Friedman, J., Hastie, T., Hoefling, H., Tibshirani, R., 2007a. Pathwise coordinate optimization. Annals of Applied Statistics 2 (1), 302–332.

Friedman, J., Hastie, T., Tibshirani, R., 2007b. Sparse inverse covariance estimation with the graphical Lasso. Biostatistics.

Friedman, J., T. Hastie, T., Tibshirani, R., 2010. A note on the group Lasso and a sparse group Lasso. Tech. Rep. arXiv:1001.0736v1, ArXiv.

Friston, K., Holmes, A., Worsley, K., Poline, J.-P., Frith, C., Frackowiak, R., 1995. Statistical parametric maps in functional imaging: A general linear approach. Human brain mapping 2 (4), 189–210.

Fu, W., 1998. Penalized regressions: The bridge vs. the Lasso. Journal of Computational and Graphical Statistics 7 (3).

Fuchs, J., 2005. Recovery of exact sparse representations in the presence of bounded noise. IEEE Trans. Information Theory 51 (10), 3601–3608.

Garey, M., Johnson, D. S., 1979. Computers and intractability: A Guide to the theory of NP-completeness. Freeman, San Francisco.

Garnaev, A., Gluskin, E., 1984. The widths of a Euclidean ball. Dokl. Akad. Nauk USSR 277, 1048–1052, english transl. Soviet Math. Dokl. 30 (200–204).

Garrigues, P., Olshausen, B., 2010. Group sparse coding with a Laplacian scale mixture prior. In: Proc. of Neural Information Processing Systems (NIPS). pp. 676–684.

Gaussier, E., Goutte, C., 2005. Relation between PLSA and NMF and implications. In: Proc. of the 28th annual international ACM SIGIR conference on Research and development in information retrieval. ACM, pp. 601–602.

Gilbert, A., Hemenway, B., Rudra, A., Strauss, M., Wootters, M., 2012. Recovering simple signals. In: Information Theory and Applications Workshop (ITA). pp. 382–391.

Gilbert, A., Strauss, M., 2007. Group testing in statistical signal recovery. Technometrics 49 (3), 346–356.

Goldstein, R., Alia-Klein, N., Tomasi, D., Honorio, J., Maloney, T., Woicik, P., Wang, R., Telang, F., Volkow, N., 2009. Anterior cingulate cortex hypoactivations to an emotionally salient task in cocaine addiction. Proceedings of the National Academy of Sciences, USA.

Gorodnitsky, F., Rao, B., 1997. Sparse signal reconstruction from limited data using FOCUSS: A reweighted norm minimization algorithm. IEEE Trans. Signal Proc. 45, 600–616.

Greenshtein, E., Ritov, Y., 2004. Persistence in high-dimensional linear predictor selection and the virtue of overparametrization. Bernoulli 10 (6), 971–988.

Gregor, K., LeCun, Y., 2010. Learning fast approximations of sparse coding. In: Proc. of the 27th International Conference on Machine Learning (ICML). pp. 399–406.

Gribonval, R., Nielsen, M., 2003. Sparse representations in unions of bases. IEEE Trans. Inform. Theory 49 (12), 3320–3325.

Grimmett, G. R., 1973. A theorem about random fields. Bulletin of the London Mathematical Society 5 (1), 81–84.

Guillamet, D., Vitrià, J., 2002. Non-negative matrix factorization for face recognition. In: Topics in Artificial Intelligence. Springer, pp. 336–344.

Hammersley, J., Clifford, P., 1971. Markov fields on finite graphs and lattices. unpublished manuscript.

Hastie, T., Tibshirani, R., Friedman, J., 2009. The elements of statistical learning: data mining, inference, and prediction, 2nd edition. New York: Springer-Verlag.

Haxby, J., Gobbini, M., Furey, M., Ishai, A., Schouten, J., Pietrini, P., 2001. Distributed and overlapping representations of faces and objects in ventral temporal cortex. Science 293 (5539), 2425–2430.

Heckerman, D., 1995. A tutorial on learning Bayesian networks. Tech. Rep. Tech. Report MSR-TR-95-06, Microsoft Research.

Ho, J., Xie, Y., Vemuri, B., 2013. On a nonlinear generalization of sparse coding and dictionary learning. In: Proc. of the International Conference on Machine Learning (ICML). pp. 1480–1488.

Ho, N.-D., 2008. Nonnegative matrix factorization algorithms and applications. Ph.D. thesis, École Polytechnique.

Hoerl, A., Kennard, R., 1988. Ridge regression. Encyclopedia of Statistical Sciences 8 (2), 129–136.

Honorio, J., Jaakkola, T., 2013. Inverse covariance estimation for high-dimensional data in linear time and space: Spectral methods for Riccati and sparse models. In: Proc. of Uncertainty in Artificial Intelligence (UAI).

Honorio, J., Ortiz, L., Samaras, D., Paragios, N., Goldstein, R., 2009. Sparse and locally constant Gaussian graphical models. In: Proc. of Neural Information Processing Systems (NIPS). pp. 745–753.

Honorio, J., Samaras, D., Rish, I., Cecchi, G., 2012. Variable selection for Gaussian graphical models. In: Proc. of the International Conference on Artificial Intelligence and Statistics (AISTATS). pp. 538–546.

Hoyer, P., 2004. Non-negative matrix factorization with sparseness constraints. Journal of Machine Learning Research 5, 1457–1469.

Hsieh, C.-J., Sustik, M., Dhillon, I., Ravikumar, P., Poldrack, R., 2013. BIG & QUIC: Sparse inverse covariance estimation for a million variables. In: Proc. of Neural Information Processing Systems (NIPS). pp. 3165–3173.

Huang, S., Li, J., Sun, L., Liu, J., Wu, T., Chen, K., Fleisher, A., Reiman, E., Ye, J., 2009. Learning brain connectivity of Alzheimer's disease from neuroimaging data. In: Proc. of Neural Information Processing Systems (NIPS). Vol. 22. pp. 808–816.

Huang, S., Li, J., Ye, J., Fleisher, A., Chen, K., Wu, T., Reiman, E., 2013. The Alzheimer's Disease Neuroimaging Initiative, 2013. A sparse structure learning algorithm for gaussian bayesian network identification from high-dimensional data. IEEE Transactions on Pattern Analysis and Machine Intelligence 35 (6), 1328–1342.

Huber, P., 1964. Robust estimation of a location parameter. The Annals of Mathematical Statistics 35 (1), 73–101.

Ishwaran, H., Rao, J., 2005. Spike and slab variable selection: Frequentist and Bayesian strategies. Ann. Statist. 33 (2), 730–773.

Jacob, L., Obozinski, G., Vert, J.-P., 2009. Group Lasso with overlap and graph Lasso. In: Proc. of the 26th Annual International Conference on Machine Learning (ICML). pp. 433–440.

Jalali, A., Chen, Y., Sanghavi, S., Xu, H., 2011. Clustering partially observed graphs via convex optimization. In: Proc. of the 28th International Conference on Machine Learning (ICML). pp. 1001–1008.

James, G. M., Radchenko, P., Lv, J., 2009. DASSO: Connections between the Dantzig selector and Lasso. Journal of The Royal Statistical Society Series B 71 (1), 127–142.

Jenatton, R., 2011. Structured sparsity-inducing norms: Statistical and algorithmic properties with applications to neuroimaging. Ph.D. thesis, Ecole Normale Superieure de Cachan.

Jenatton, R., Audibert, J.-Y., Bach, F., 2011a. Structured variable selection with sparsity-inducing norms. Journal of Machine Learning Research 12, 2777–2824.

Jenatton, R., Gramfort, A., Michel, V., Obozinski, G., Bach, F., Thirion, B., 2011b. Multi-scale mining of fMRI data with hierarchical structured sparsity. In: Proceedings of the 2011 International Workshop on Pattern Recognition in NeuroImaging (PRNI). IEEE, pp. 69–72.

Jenatton, R., Mairal, J., Obozinski, G., Bach, F., 2011c. Proximal methods for hierarchical sparse coding. Journal of Machine Learning Research 12, 2297–2334.

Jenatton, R., Obozinski, G., Bach, F., 2010. Structured sparse principal component analysis. In: Proc. of International Conference on Artificial Intelligence and Statistics. pp. 366–373.

Ji, S., Xue, Y., Carin, L., June 2008. Bayesian compressive sensing. IEEE Trans. on Signal Processing 56 (6), 2346–2356.

Johnson, W., Lindenstrauss, J., 1984. Extensions of Lipschitz mappings into a Hilbert space. In: Conf. in Modern Analysis and Probability. Vol. 26. Amer. Math. Soc., Providence, RI, pp. 189–206.

Jolliffe, I., Trendafilov, N., Uddin, M., 2003. A modified principal component technique based on the Lasso. Journal of Computational and Graphical Statistics 12, 531–547.

Jordan, M., 2000. Graphical models. Statistical Science (Special Issue on Bayesian Statistics) 19, 140–155.

Kakade, S., Shamir, O., Sindharan, K., Tewari, A., 2010. Learning exponential families in high-dimensions: Strong convexity and sparsity. In: Proc. of the International Conference on Artificial Intelligence and Statistics (AISTATS). pp. 381–388.

Kambadur, P., Lozano, A. C., 2013. A parallel, block greedy method for sparse inverse covariance estimation for ultra-high dimensions. In: Proc. of the International Conference on Artificial Intelligence and Statistics (AISTATS). pp. 351–359.

Kannan, R., Lovász, L., Simonovits, M., 1997. Random walks and an O*(n5) volume algorithm for convex bodies. Random structures and algorithms 11 (1), 1–50.

Kasiviswanathan, S. P., Wang, H., Banerjee, A., Melville, P., 2012. Online L1-dictionary learning with application to novel document detection. In: Proc. of Neural Information Processing Systems (NIPS). pp. 2267–2275.

Kim, H., Park, H., 2007. Sparse non-negative matrix factorizations via alternating non-negativity-constrained least squares for microarray data analysis. Bioinformatics 23 (12), 1495–1502.

Kim, S., Xing, E., 2010. Tree-guided group Lasso for multi-task regression with structured sparsity. In: Proc. of the 27th International Conference on Machine Learning (ICML). pp. 543–550.

Kim, T., Shakhnarovich, G., Urtasun, R., 2010. Sparse coding for learning interpretable spatio-temporal primitives. In: Proc. of Neural Information Processing Systems (NIPS). pp. 1117–1125.

Kimeldorf, G., Wahba, G., 1971. Some results on Tchebycheffian spline function. J. Math. Anal. Applications 33, 82–95.

Kindermann, R., Snell, J., 1980. Markov Random Fields and Their Applications. American Mathematical Society.

Knight, K., Fu, W., 2000. Asymptotics for Lasso-type estimators. Ann. Statist. 28 (5), 1356–1378.

Koller, D., Friedman, N., 2009. Probabilistic Graphical Models: Principles and Techniques. MIT Press.

Koltchinskii, V., 2009. The Dantzig selector and sparsity oracle inequalities. Bernoulli 15 (3), 799–828.

Kotelnikov, V., 1933. On the transmission capacity of ether and wire in electric communications. Izd. Red. Upr. Svyazi RKKA (in Russian).

Kotelnikov, V., 2006. On the transmission capacity of ether and wire in electric communications (1933). Physics-Uspekhi 49 (7), 736–744.

Kreutz-Delgado, K., Murray, J., Rao, B., Engan, K., Lee, T.-W., Sejnowski, T., 2003. Dictionary learning algorithms for sparse representation. Neural Computation 15 (2), 349–396.

Kruskal, J., 1977. Three-way arrays: Rank and uniqueness of trilinear decompositions, with application to arithmetic complexity and statistics. Linear algebra and its applications 18 (2), 95–138.

Lanckriet, G., Cristianini, N., Bartlett, P., Ghaoui, L. E., Jordan, M., 2004. Learning the kernel matrix with semidefinite programming. Journal of Machine Learning Research 5, 27–72.

Lauritzen, S., 1996. Graphical Models. Oxford University Press.

Lawton, W., Sylvestre, E., 1971. Self modeling curve resolution. Technometrics 13 (3), 617–633.

Ledoux, M., Talagrand, M., 2011. Probability in Banach spaces: Isoperimetry and processes. Vol. 23. Springer.

Lee, H., Battle, A., Raina, R., Ng, A., 2006a. Efficient sparse coding algorithms. In: Proc. of Neural Information Processing Systems (NIPS). pp. 801–808.

Lee, J., Telang, F., Springer, C., Volkow, N., 2003. Abnormal brain activation to visual stimulation in cocaine abusers. Life Sciences.

Lee, S.-I., Ganapathi, V., Koller, D., 2006b. Efficient structure learning of Markov networks using l_1-regularization. In: Proc. of Neural Information Processing Systems (NIPS). pp. 817–824.

Lesage, S., Gribonval, R., Bimbot, F., Benaroya, L., 2005. Learning unions of orthonormal bases with thresholded singular value decomposition. In: Proc. of the International Conference on Acoustics, Speech, and Signal Processing (ICASSP). Vol. 5. pp. 293–296.

Lewicki, M., Olshausen, B., 1999. A probabilistic framework for the adaptation and comparison of image codes. Journal of the Optical Society of America A: Optics, Image Science and Vision 16 (7), 1587–1601.

Lewicki, M., Sejnowski, T., 2000. Learning overcomplete representations. Neural Computation 12, 337–365.

Li, J., Tao, D., 2010. Simple exponential family PCA. In: Proc. of International Conference on Artificial Intelligence and Statistics (AISTATS). pp. 453–460.

Li, S. Z., Hou, X., Zhang, H., Cheng, Q., 2001. Learning spatially localized, parts-based representation. In: Proc. of the 2001 IEEE Computer Society Conference on Computer Vision and Pattern Recognition (CVPR). Vol. 1. IEEE, pp. I–207.

Lin, Y., Zhang, H., 2006. Component selection and smoothing in smoothing spline analysis of variance models. Annals of Statistics 34, 2272–2297.

Lin, Y., Zhu, S., Lee, D., Taskar, B., 2009. Learning sparse Markov network structure via ensemble-of-trees models. In: Proc. of the International Conference on Artificial Intelligence and Statistics (AISTATS). pp. 360–367.

Lions, P., Mercier, B., 1979. Splitting algorithms for the sum of two nonlinear operators. SIAM Journal on Numerical Analysis 16 (6), 964–979.

Liu, H., Palatucci, M., Zhang, J., 2009a. Blockwise coordinate descent procedures for the multi-task Lasso, with applications to neural semantic basis discovery. In: Proc. of the 26th Annual International Conference on Machine Learning (ICML). ACM, pp. 649–656.

Liu, H., Zhang, J., 2009. Estimation consistency of the group Lasso and its applications. Journal of Machine Learning Research - Proceedings Track 5, 376–383.

Liu, J., Ji, S., Ye, J., 2009b. Multi-task feature learning via efficient $l_{2,1}$-norm minimization. In: Proc. of the Twenty-Fifth Conference on Uncertainty in Artificial Intelligence (UAI). AUAI Press, pp. 339–348.

Lozano, A., Abe, N., Liu, Y., Rosset, S., 2009a. Grouped graphical Granger modeling methods for temporal causal modeling. In: Proc. of the 15th ACM SIGKDD International Conference on Knowledge Discovery and Data Mining. pp. 577–586.

Lozano, A., Swirszcz, G., Abe, N., 2009b. Grouped orthogonal matching pursuit for variable selection and prediction. In: Proc. of Neural Information Processing Systems (NIPS). pp. 1150–1158.

Lu, Z., 2009. Smooth optimization approach for sparse covariance selection. SIAM Journal on Optimization 19 (4), 1807–1827.

Lv, J., Fan, Y., 2009. A unified approach to model selection and sparse recovery using regularized least squares. The Annals of Statistics 37 (6A), 3498–3528.

Mairal, J., Bach, F., Ponce, J., Sapiro, G., 2009. Online dictionary learning for sparse coding. In: Proc. of the 26th Annual International Conference on Machine Learning (ICML). ACM, pp. 689–696.

Mairal, J., Bach, F., Ponce, J., Sapiro, G., 2010. Online learning for matrix factorization and sparse coding. Journal of Machine Learning Research 11, 19–60.

Mairal, J., Yu, B., 2012. Complexity analysis of the Lasso regularization path. In: Proc. of the 29th International Conference on Machine Learning (ICML). pp. 353–360.

Malioutov, D., Çetin, M., Willsky, A., 2005. A sparse signal reconstruction perspective for source localization with sensor arrays. IEEE Transactions on Signal Processing 53 (8), 3010–3022.

Mallat, S., Davis, G., Zhang, Z., 1994. Adaptive time-frequency decompositions. SPIE Journal of Optical Engineering 33, 2183–2191.

Mallat, S., Zhang, Z., 1993. Matching pursuits with time-frequency dictionaries. IEEE Transactions on Signal Processing 41, 3397–3415.

Marlin, B., Murphy, K., 2009. Sparse Gaussian graphical models with unknown block structure. In: Proc. of the 26th Annual International Conference on Machine Learning (ICML). ACM, pp. 705–712.

Martinet, B., 1970. Régularisation d-inéquations variationnelles par approximations successives. Revue franaise d-informatique et de recherche opérationnelle, série rouge.

Matoušek, J., 2002. Lectures on discrete geometry. Springer Verlag.

Maurer, A., Pontil, M., Romera-Paredes, B., 2013. Sparse coding for multitask and transfer learning. In: Proc. of International Conference on Machine Learning (ICML). pp. 343–351.

Mccullagh, P., Nelder, J., 1989. Generalized Linear Models, 2nd ed. Chapman and Hall, London.

Meier, L., van de Geer, S., Bühlmann, P., 2008. The group Lasso for logistic regression. J. Royal Statistical Society: Series B 70 (1), 53–71.

Meinshausen, N., 2007. Relaxed Lasso. Computational Statistics and Data Analysis 52 (1), 374–293.

Meinshausen, N., Bühlmann, P., 2006. High dimensional graphs and variable selection with the Lasso. Annals of Statistics 34(3), 1436–1462.

Meinshausen, N., Bühlmann, P., 2010. Stability selection. Journal of the Royal Statistical Society: Series B (Statistical Methodology) 72 (4), 417–473.

Meinshausen, N., Rocha, G., Yu, B., 2007. Discussion: A tale of three cousins: Lasso, L2Boosting and Dantzig. The Annals of Statistics, 2373–2384.

Milman, V., Pajor, A., 1989. Isotropic position and inertia ellipsoids and zonoids of the unit ball of a normed n-dimensional space. Geometric aspects of functional analysis, 64–104.

Milman, V., Schechtman, G., 1986. Asymptotic theory of finite dimensional normed spaces. Springer Verlag.

Mishra, B., Meyer, G., Bach, F., Sepulchre, R., 2013. Low-rank optimization with trace norm penalty. SIAM Journal on Optimization 23 (4), 2124–2149.

Mitchell, T., Hutchinson, R., Niculescu, R., Pereira, F., Wang, X., Just, M., Newman, S., 2004. Learning to decode cognitive states from brain images. Machine Learning 57, 145–175.

Moghaddam, B., Weiss, Y., Avidan, S., 2006. Generalized spectral bounds for sparse LDA. In: Proc. of the 23rd International Conference on Machine Learning (ICML). ACM, pp. 641–648.

Moreau, J., 1962. Fonctions convexes duales et points proximaux dans un espace hilbertien. Comptes-Rendus de l-Académie des Sciences de Paris, Série A, Mathèmatiques 255, 2897–2899.

Morioka, N., Shiníchi, S., 2011. Generalized Lasso based approximation of sparse coding for visual recognition. In: Proc. of Neural Information Processing Systems (NIPS). pp. 181–189.

Moussouris, J., 1974. Gibbs and Markov systems with constraints. Journal of statistical physics 10, 11–33.

Muthukrishnan, S., 2005. Data streams: Algorithms and applications. Now Publishers Inc.

Nardi, Y., Rinaldo, A., 2008. On the asymptotic properties of the group Lasso estimator for linear models. Electronic Journal of Statistics 2, 605–633.

Natarajan, K., 1995. Sparse approximate solutions to linear systems. SIAM J. Comput. 24, 227–234.

Needell, D., Tropp, J. A., 2008. Iterative signal recovery from incomplete and inaccurate samples. Appl. Comput. Harmon. Anal. 26, 301–321.

Needell, D., Vershynin, R., 2009. Uniform uncertainty principle and signal recovery via regularized orthogonal matching pursuit. Foundations of Computational Mathematics 9, 317–334.

Negahban, S., Ravikumar, P., Wainwright, M., Yu, B., 2009. A unified framework for high-dimensional analysis of M-estimators with decomposable regularizers. In: Proc. of Neural Information Processing Systems (NIPS). pp. 1348–1356.

Negahban, S., Ravikumar, P., Wainwright, M., Yu, B., 2012. A unified framework for high-dimensional analysis of M-estimators with decomposable regularizers. Statistical Science 27 (4), 438–557.

Negahban, S., Wainwright, M., 2011. Estimation of (near) low-rank matrices with noise and high-dimensional scaling. The Annals of Statistics 39 (2), 1069–1097.

Nemirovsky, A., Yudin, D., 1983. Problem Complexity and Method Efficiency in Optimization. Wiley-Interscience Series in Discrete Mathematics, John Wiley & Sons, New York.

Nesterov, Y., 1983. A method for solving the convex programming problem with convergence rate $o(1/k^2)$ (in Russian). Dokl. Akad. Nauk SSSR 269, 543–547.

Nesterov, Y., 2005. Smooth minimization of non-smooth functions. Mathematical programming 103 (1), 127–152.

Nocedal, J., Wright, S., 2006. Numerical Optimization, Second Edition. Springer.

Nowak, R., Figueiredo, M., 2001. Fast wavelet-based image deconvolution using the EM algorithm. In: Proc. 35th Asilomar Conf. on Signals, Systems, and Computers. Vol. 1. pp. 371–375.

Nyquist, H., 1928. Certain topics in telegraph transmission theory. Transactions of the AIEE 47, 617–644.

Obozinski, G., Jacob, L., Vert, J.-P., 2011. Group Lasso with overlaps: The latent group Lasso approach. Tech. Rep. 1110.0413, arXiv.

Obozinski, G., Taskar, B., Jordan, M., 2010. Joint covariate selection and joint subspace selection for multiple classification problems. Statistics and Computing 20 (2), 231–252.

Olsen, P. A., Öztoprak, F., Nocedal, J., Rennie, S., 2012. Newton-like methods for sparse inverse covariance estimation. In: Proc. of Neural Information Processing Systems (NIPS). pp. 764–772.

Olshausen, B., Field, D., 1996. Emergence of simple-cell receptive field properties by learning a sparse code for natural images. Nature 381 (6583), 607–609.

Olshausen, B., Field, D., 1997. Sparse coding with an overcomplete basis set: A strategy employed by V1? Vision Research 37, 3311–3325.

Osborne, M., Presnell, B., Turlach, B., 2000a. A new approach to variable selection in least squares problems. IMA Journal of Numerical Analysis 20 (3), 389–403.

Osborne, M., Presnell, B., Turlach, B., 2000b. On the Lasso and its dual. Journal of Computational and Graphical Statistics 9 (2), 319–337.

Park, M., Hastie, T., 2007. An $l1$ regularization-path algorithm for generalized linear models. JRSSB 69 (4), 659–677.

Patel, V., Chellappa, R., 2013. Sparse Representations and Compressive Sensing for Imaging and Vision. Springer Briefs in Electrical and Computer Engineering.

Pati, Y., Rezaiifar, R., Krishnaprasad, P., November 1993. Orthogonal matching pursuit: Recursive function approximation with applications to wavelet decomposition. In: Proc. 27th Annu. Asilomar Conf. Signals, Systems, and Computers. Vol. 1. pp. 40–44.

Pearl, J., 1988. Probabilistic reasoning in intelligent systems: Networks of plausible inference. Morgan Kaufmann, San Mateo, California.

Pearl, J., 2000. Causality: Models, Reasoning and Inference. Cambridge University Press.

Pearl, J., Paz, A., 1987. Graphoids: A graph based logic for reasoning about relevance relations. Advances in Artificial Intelligence II, 357–363.

Pearson, K., 1901. On lines and planes of closest fit to systems of points in space. Philosophical Magazine 2 (11), 559–572.

Pittsburgh EBC Group, 2007. PBAIC Homepage: http://www.ebc.pitt.edu/2007/competition.html.

Preston, C. J., 1973. Generalized Gibbs states and Markov random fields. Advances in Applied Probability 5 (2), 242–261.

Quattoni, A., Carreras, X., Collins, M., Darrell, T., 2009. An efficient projection for $l_{1,\infty}$ regularization. In: Proc. of the 26th Annual International Conference on Machine Learning (ICML). pp. 857–864.

Rauhut, H., 2008. Stability results for random sampling of sparse trigonometric polynomials. IEEE Transactions on Information Theory 54 (12), 5661–5670.

Ravikumar, P., Raskutti, G., Wainwright, M., Yu, B., 2009. Model selection in Gaussian graphical models: High-dimensional consistency of l_1-regularized MLE. In: Proc. of Neural Information Processing Systems (NIPS). pp. 1329–1336.

Ravikumar, P., Wainwright, M., Lafferty, J., 2010. High-dimensional Ising model selection using l_1-regularized logistic regression. Ann. Statist. 38, 1287–1319.

Recht, B., Fazel, M., Parrilo, P., 2010. Guaranteed minimum-rank solutions of linear matrix equations via nuclear norm minimization. SIAM Review 52 (3), 471–501.

Resources, C. S., 2010. http://dsp.rice.edu/cs.

Rish, I., Brodie, M., Ma, S., Odintsova, N., Beygelzimer, A., Grabarnik, G., Hernandez, K., 2005. Adaptive diagnosis in distributed systems. IEEE Transactions on Neural Networks (special issue on Adaptive Learning Systems in Communication Networks) 16 (5), 1088–1109.

Rish, I., Cecchi, G., Baliki, M., Apkarian, A., 2010. Sparse regression models of pain perception. In: Brain Informatics. Springer, pp. 212–223.

Rish, I., Cecchi, G., Thyreau, B., Thirion, B., Plaze, M., Paillere-Martinot, M., Martelli, C., Martinot, J.L., Poline, J.B. 2013. Schizophrenia as a network disease: Disruption of emergent brain function in patients with auditory hallucinations. PLoS ONE 8 (1).

Rish, I., Cecchi, G. A., Heuton, K., February 2012a. Schizophrenia classification using fMRI-based functional network features. In: Proc. of SPIE Medical Imaging.

Rish, I., Cecchi, G. A., Heuton, K., Baliki, M. N., Apkarian, A. V., February 2012b. Sparse regression analysis of task-relevant information distribution in the brain. In: Proc. of SPIE Medical Imaging.

Rish, I., Grabarnik, G., 2009. Sparse signal recovery with exponential-family noise. In: Proc. of the 47th Annual Allerton Conference on Communication, Control, and Computing. pp. 60–66.

Rish, I., Grabarnik, G., Cecchi, G., Pereira, F., Gordon, G., 2008. Closed-form supervised dimensionality reduction with generalized linear models. In: Proc. of the 25th International Conference on Machine Learning (ICML). ACM, pp. 832–839.

Ritov, Y., December 2007. Discussion: The Dantzig selector: Statistical estimation when p is much larger than n. The Annals of Statistics 35 (6), 2370–2372.

Robertson, H., 1940. Communicated by s. goldstein received 15 november 1939 the statistical theory of isotropic turbulence, initiated by taylor (3) and extended by de karman and howarth (2), has proved of value in attacking problems. In: Proceedings of the Cambridge Philosophical Society: Mathematical and Physical Sciences. Vol. 36. Cambridge University Press, p. 209.

Rockafeller, R., 1970. Convex Analysis. Princeton University Press.

Rohde, A., Tsybakov, A., April 2011. Estimation of high-dimensional low-rank matrices. The Annals of Statistics 39 (2), 887–930.

Rosset, S., Zhu, J., 2007. Piecewise linear regularized solution paths. Annals of Statistics 35 (3).

Roth, V., Fischer, B., 2008. The group Lasso for generalized linear models: Uniqueness of solutions and efficient algorithms. In: Proc. of the 25th International Conference on Machine learning (ICML). pp. 848–855.

Rothman, A., Bickel, P., Levina, E., Zhu, J., 2008. Sparse permutation invariant covariance estimation. Electronic Journal of Statistics 2, 494–515.

Rudelson, M., 1999. Random vectors in the isotropic position. Journal of Functional Analysis 164 (1), 60–72.

Rudelson, M., 2007. Probabilistic and combinatorial methods in analysis, cbms lecture notes, preprint.

Rudelson, M., Vershynin, R., 2006. Sparse reconstruction by convex relaxation: Fourier and Gaussian measurements. In: Proc. of the 40th Annual Conference on Information Sciences and Systems. pp. 207–212.

Rudelson, M., Vershynin, R., 2008. On sparse reconstruction from Fourier and Gaussian measurements. Communications on Pure and Applied Mathematics 61 (8), 1025–1045.

Rudin, L., Osher, S., Fatemi, E., 1992. Nonlinear total variation based noise removal algorithms. Physica D 60, 259–268.

Sajama, S., Orlitsky, A., 2004. Semi-parametric exponential family PCA. In: Proc. of Neural Information Processing Systems (NIPS). pp. 1177–1184.

Santosa, F., Symes, W., 1986. Linear inversion of band-limited reflection seismograms. SIAM Journal on Scientific and Statistical Computing 7 (4), 1307–1330.

Scheinberg, K., Asadi, N. B., Rish, I., 2009. Sparse MRF learning with priors on regularization parameters. Tech. Rep. RC24812, IBM T.J. Watson Research Center.

Scheinberg, K., Ma, S., 2011. Optimization methods for sparse inverse covariance selection. In: Sra, S., Nowozin, S., Wright, S. J. (Eds.), Optimization for Machine Learning. MIT Press.

Scheinberg, K., Ma, S., Goldfarb, D., 2010a. Sparse inverse covariance selection via alternating linearization methods. In: Proc. of Neural Information Processing Systems (NIPS). pp. 2101–2109.

Scheinberg, K., Rish, I., 2010. Learning sparse Gaussian Markov networks using a greedy coordinate ascent approach. In: Machine Learning and Knowledge Discovery in Databases. Springer, pp. 196–212.

Scheinberg, K., Rish, I., Asadi, N. B., January 2010b. Sparse Markov net learning with priors on regularization parameters. In: Proc. of International Symposium on AI and Mathematics (AIMATH 2010).

Schmidt, M., 2010. Graphical Model Structure Learning using L1-Regularization. Ph.D. thesis, University of British Columbia.

Schmidt, M., Berg, E. V. D., Friedl, M., Murphy, K., 2009. Optimizing costly functions with simple constraints: A limited-memory projected quasi-Newton algorithm. In: Proc. of the International Conference on Artificial Intelligence and Statistics (AISTATS). pp. 456–463.

Schmidt, M., Murphy, K., 2010. Convex structure learning in log-linear models: Beyond pairwise potentials. In: Proc. of the International Conference on Artificial Intelligence and Statistics (AISTATS). pp. 709–716.

Schmidt, M., Niculescu-Mizil, A., Murphy, K., 2007. Learning graphical model structure using l_1-regularization paths. In: Proc. of the International Conference on Artificial Intelligence (AAAI). Vol. 7. pp. 1278–1283.

Schmidt, M., Rosales, R., Murphy, K., Fung, G., 2008. Structure learning in random fields for heart motion abnormality detection. In: Proc. of the IEEE Conference on Computer Vision and Pattern Recognition (CVPR). IEEE, pp. 1–8.

Schrijver, A., 1986. Theory of linear and integer programming. John Wiley & Sons, Inc., New York, NY, USA.

Shannon, C. E., January 1949. Communication in the presence of noise. Proc. Institute of Radio Engineers 37 (1), 10–21.

Shashua, A., Hazan, T., 2005. Non-negative tensor factorization with applications to statistics and computer vision. In: Proc. of the 22nd International Conference on Machine Learning (ICML). ACM, pp. 792–799.

Shawe-Taylor, J., Cristianini, N., 2004. Kernel Methods for Pattern Analysis. Cambridge University Press.

Shelton, J., Sterne, P., Bornschein, J., Sheikh, A.-S., Lücke, J., 2012. Why MCA? Nonlinear sparse coding with spike-and-slab prior for neurally plausible image encoding. In: Proc. of Neural Information Processing Systems (NIPS). pp. 2285–2293.

Shepp, L., Logan, B., 1974. The fourier reconstruction of a head section. IEEE Transactions on Nuclear Science 21 (3), 21–43.

Sherman, S., 1973. Markov random fields and Gibbs random fields. Israel Journal of Mathematics 14 (1), 92–103.

Sjöstrand, K. 2005. Matlab implementation of LASSO, LARS, the elastic net and SPCA: http://www2.imm.dtu.dk/pubdb/views/publication_details.php?id=3897.

Skretting, K., Engan, K., 2010. Recursive least squares dictionary learning algorithm. IEEE Transactions on Signal Processing 58 (4), 2121–2130.

Srebro, N., Rennie, J., Jaakkola, T., 2004. Maximum-margin matrix factorization. In: Proc. of Neural Information Processing Systems (NIPS). Vol. 17. pp. 1329–1336.

Starck, J.-L., Donoho, D., Candès, E., 2003a. Astronomical image representation by the curvelet transform. Astronomy and Astrophysics 398 (2), 785–800.

Starck, J.-L., Nguyen, M., Murtagh, F., 2003b. Wavelets and curvelets for image deconvolution: A combined approach. Signal Processing 83, 2279–2283.

Strohmer, T., Heath, R., 2003. Grassmannian frames with applications to coding and communication. Applied and Computational Harmonic Analysis 14 (3), 257–275.

Sun, L., Patel, R., Liu, J., Chen, K., Wu, T., Li, J., Reiman, E., Ye, J., 2009. Mining brain region connectivity for Alzheimer's disease study via sparse inverse covariance estimation. In: Proc. of the 15th ACM SIGKDD International Conference on Knowledge Discovery and Data Mining (KDD). ACM, pp. 1335–1344.

Szlam, A., Gregor, K., Cun, Y., 2011. Structured sparse coding via lateral inhibition. In: Proc. of Neural Information Processing Systems (NIPS). pp. 1116–1124.

Talagrand, M., 1996. Majorizing measures: The generic chaining. The Annals of Probability 24 (3), 1049–1103.

Tibshirani, R., 1996. Regression shrinkage and selection via the Lasso. Journal of the Royal Statistical Society, Series B 58 (1), 267–288.

Tibshirani, R., 2013. The Lasso problem and uniqueness. Electronic Journal of Statistics 7, 1456–1490.

Tibshirani, R., Saunders, M., Rosset, S., Zhu, J., Knight, K., 2005. Sparsity and smoothness via the fused Lasso. Journal of the Royal Statistical Society Series B, 91–108.

Tibshirani, R., Wang, P., 2008. Spatial smoothing and hot spot detection for CGH data using the fused Lasso. Biostatistics 9 (1), 18–29.

Tipping, M., 2001. Sparse Bayesian learning and the Relevance Vector Machine. Journal of Machine Learning Research 1, 211–244.

Tipping, M., Bishop, C., 1999. Probabilistic principal component analysis. Journal of the Royal Statistical Society, Series B 21 (3), 611–622.

Toh, K.-C., Yun, S., 2010. An accelerated proximal gradient algorithm for nuclear norm regularized least squares problems. Pacific J. Optim. 6, 615–640.

Tosic, I., Frossard, P., 2011. Dictionary learning: What is the right representation for my signal? IEEE Signal Proc. Magazine 28 (2), 27–38.

Tropp, A., 2006. Just relax: Convex programming methods for subset slection and sparse approximation. IEEE Trans. Inform. Theory 51 (3), 1030–1051.

Tseng, P., Yun, S., 2009. A coordinate gradient descent method for nonsmooth separable minimization. Mathematical Programming 117 (1), 387–423.

Turlach, B., Venables, W., Wright, S., 2005. Simultaneous variable selection. Technometrics 47 (3), 349–363.

van de Geer, S., 2008. High-dimensional generalized linear models and the Lasso. Ann. Statist. 36, 614–645.

Vandenberghe, L., Boyd, S., Wu, S., 1998. Determinant maximization with linear matrix inequality constraints. SIAM J. Matrix Anal. Appl. 19 (2), 499–533.

Vardi, Y., 1996. Network tomography: Estimating source-destination traffic intensities from link data. J. Amer. Statist. Assoc. 91, 365–377.

Vershynin, R., 2012. Introduction to the non-asymptotic analysis of random matrices. In: Eldar, Y., Kutyniok, G. (Eds.), Compressed Sensing, Theory and Application. Cambridge University Press, pp. 210–268.

Wainwright, M., May 2009. Sharp thresholds for noisy and high-dimensional recovery of sparsity using l_1-constrained quadratic programming (Lasso). IEEE Transactions on Information Theory 55, 2183–2202.

Wainwright, M., Ravikumar, P., Lafferty, J., 2007. High-dimensional graphical model selection using l_1-regularized logistic regression. Proc. of Neural Information Processing Systems (NIPS) 19, 1465–1472.

Weisberg, S., 1980. Applied Linear Regression. Wiley, New York.

Welch, L., 1974. Lower bounds on the maximum cross correlation of signals (corresp.). IEEE Transactions on Information Theory 20 (3), 397–399.

Whittaker, E., 1915. On the functions which are represented by the expansion of interpolating theory. Proc. R. Soc. Edinburgh 35, 181–194.

Whittaker, J., 1929. The Fourier theory of the cardinal functions. Proc. Math. Soc. Edinburgh 1, 169–176.

Whittaker, J., 1990. Graphical Models in Applied Multivariate Statistics. Wiley.

Wipf, D., Rao, B., August 2004. Sparse Bayesian learning for basis selection. IEEE Transactions on Signal Processing 52 (8), 2153–2164.

Witten, D., Tibshirani, R., Hastie, T., 2009. A penalized matrix decomposition, with applications to sparse canonical correlation analysis and principal components. Biostatistics 10 (3), 515–534.

Xiang, J., Kim, S., 2013. A* Lasso for learning a sparse Bayesian network structure for continuous variables. In: Proceedings of Neural Information Processing Systems (NIPS). pp. 2418–2426.

Xiang, Z., Xu, H., Ramadge, P., 2011. Learning sparse representations of high dimensional data on large scale dictionaries. In: Proc. of Neural Information Processing Systems (NIPS). Vol. 24. pp. 900–908.

Xu, W., Liu, X., Gong, Y., 2003. Document clustering based on non-negative matrix factorization. In: Proc. of the 26th Annual International ACM SIGIR Conference on Research and Development in Information Retrieval. SIGIR '03. ACM, pp. 267–273.

Yaghoobi, M., Blumensath, T., Davies, M., 2009. Dictionary learning for sparse approximations with the majorization method. IEEE Transactions on Signal Processing 57 (6), 2178–2191.

Yuan, M., 2010. Sparse inverse covariance matrix estimation via linear programming. Journal of Machine Learning Research 11, 2261–2286.

Yuan, M., Ekici, A., Lu, Z., Monteiro, R., 2007. Dimension reduction and coefficient estimation in multivariate linear regression. Journal of the Royal Statistical Society. Series B (Methodological) 69 (3), 329–346.

Yuan, M., Lin, Y., 2006. Model selection and estimation in regression with grouped variables. Journal of the Royal Statistical Society, Series B 68, 49–67.

Yuan, M., Lin, Y., 2007. Model selection and estimation in the Gaussian graphical model. Biometrika 94(1), 19–35.

Zhao, P., Rocha, G., Yu, B., 2009. Grouped and hierarchical model selection through composite absolute penalties. Annals of Statistics 37 (6A), 3468–3497.

Zhao, P., Yu, B., November 2006. On model selection consistency of Lasso. J. Machine Learning Research 7, 2541–2567.

Zhao, P., Yu, B., 2007. Stagewise Lasso. Journal of Machine Learning Research 8, 2701–2726.

Zheng, A., Rish, I., Beygelzimer, A., 2005. Efficient test selection in active diagnosis via entropy approximation. In: Proc. of the Twenty-First Conference Annual Conference on Uncertainty in Artificial Intelligence (UAI). AUAI Press, Arlington, Virginia, pp. 675–682.

Bibliography

Zhou, M., Chen, H., Ren, L., Sapiro, G., Carin, L., Paisley, J., 2009. Non-parametric Bayesian dictionary learning for sparse image representations. In: Proc. of Neural Information Processing Systems (NIPS). pp. 2295–2303.

Zou, H., 2006. The adaptive Lasso and its oracle properties. Journal of the American Statistical Association 101 (476), 1418–1429.

Zou, H., Hastie, T., 2005. Regularization and variable selection via the Elastic Net. Journal of the Royal Statistical Society, Series B 67 (2), 301–320.

Zou, H., Hastie, T., Tibshirani, R., 2006. Sparse principal component analysis. Journal of Computational and Graphical Statistics 15 (2), 262–286.

Index